Children, Young People ar

The very notions of childhood and youth are intimately connected to contemporary norms, practices and spaces of care, caring and care-giving. The provision of care is widely figured as both the primary responsibility of parents, carers and practitioners who work with children and young people, and the primary factor in shaping children and young people's development, education, socialisation, wellbeing and contentment. However, children and young people themselves are rarely figured as key actors in the provision of care. An overwhelming presumption that children and young people are to be cared for has effectively marginalised their agency and responsibilities as carers, or in relation to practices and spaces of care.

Bringing together a significant array of multidisciplinary work on children, young people and families, this collection draws together new research on the diverse lives and experiences of children and young people as carers, as cared for, and in relation to spaces and institutions of care. It is the first collection specifically devoted to the subject of care in relation to childhood and youth. As such, the book will be a key resource for academics, practitioners and students seeking leading-edge empirical and conceptual material on this topic.

John Horton is Associate Professor in Human Geography and Associate Director of the Centre for Children and Youth (CCY) at the University of Northampton, UK.

Michelle Pyer is a Senior Researcher in the Institute for Health and Wellbeing at the University of Northampton, UK.

Routledge spaces of Childhood and Youth Series
Edited by Peter Kraftl and John Horton

The *Routledge Spaces of Childhood and Youth Series* provides a forum for original, interdisciplinary and cutting edge research to explore the lives of children and young people across the social sciences and humanities. Reflecting contemporary interest in spatial processes and metaphors across several disciplines, titles within the series explore a range of ways in which concepts such as space, place, spatiality, geographical scale, movement/mobilities, networks and flows may be deployed in childhood and youth scholarship. This series provides a forum for new theoretical, empirical and methodological perspectives and ground-breaking research that reflects the wealth of research currently being undertaken. Proposals that are cross-disciplinary, comparative and/or use mixed or creative methods are particularly welcomed, as are proposals that offer critical perspectives on the role of spatial theory in understanding children and young people's lives. The series is aimed at upper-level undergraduates, research students and academics, appealing to geographers as well as the broader social sciences, arts and humanities.

For a full list of titles in this series, please visit www.routledge.com/Routledge-Spaces-of-Childhood-and-Youth-Series/book-series/RSCYS

Young Entrepreneurs in Sub-Saharan Africa
Edited by Katherine Gough and Thilde Langevang

Children, Young People and Care
Edited by John Horton and Michelle Pyer

Children, Young People and Care

Edited by John Horton and Michelle Pyer

Routledge
Taylor & Francis Group
LONDON AND NEW YORK

First published 2017 by Routledge

2 Park Square, Milton Park, Abingdon, Oxfordshire OX14 4RN
52 Vanderbilt Avenue, New York, NY 10017

Routledge is an imprint of the Taylor & Francis Group, an informa business

First issued in paperback 2018

British Library Cataloguing in Publication Data
A catalogue record for this book is available from the British Library

Library of Congress Cataloging in Publication Data
A catalog record for this book has been requested

ISBN: 978-1-138-92088-0 (hbk)
ISBN: 978-0-367-21884-3 (pbk)

Typeset in Times New Roman
by diacriTech, Chennai

Dedication: For our parents

Contents

Figures

Tables

Boxes

Contributors

Sophie Bowlby is Visiting Research Fellow in the Department of Geography and Environmental Science at the University of Reading, UK. Her research is concerned with links between urban, social, economic and gendered geographies, with particular interests in relation to care, access, mobility and shopping, questions of identity, racialised gendering and paid work. Her concepts of 'carescapes' and 'caringscapes' (developed in collaboration with Linda McKie and colleagues) are a touchstone of many chapters in this book

Annie Rose Crowley is a doctoral researcher at the University of Glasgow, UK. Her PhD explores practitioners' perspectives of working with young women in / 'at risk of' contact with the criminal justice system in Scotland. Her international background of working within social work, the voluntary sector and the Inspectorate of Prisons inform her wider interests, which include feminist perspectives on young women's pathways into, and through, the criminal justice system, including their experiences of the care system.

Carmen Dalli is Professor of Early Childhood Education (ECE), Director of the Institute for Early Childhood Studies, and Associate Dean (Research) at the Faculty of Education at Victoria University of Wellington, New Zealand. Her research combines an interest in developmental issues in the early years with a focus on early childhood policy and pedagogy. She has published widely in the field of early childhood policy, teacher professionalism, children's experiences in ECE setting and particularly for children aged under three years. She is co-editor, with E. Jayne White, of a new book series book series (Springer) entitled *Policy and Pedagogy with Under-Three Year Olds: Cross-Disciplinary Insights and Innovations*.

Caroline Day is a Senior Lecturer in the Department of Geography at the University of Portsmouth, UK specialising in the Geographies of Development, particularly in relation to children, young people and their families. She completed her doctorate at the University of Reading, UK focusing on the transitions to adulthood of young people with and without caring responsibilities for sick or disabled parents or relatives in Zambia. She has also held research posts at the UK charities Barnardo's and Centrepoint working with vulnerable children, young people and families from a variety of social, cultural and economic backgrounds. She has conducted research in a number of sub-Saharan African countries, including Uganda, Kenya and Zambia. Her research interests focus on children, youth and families (particularly in Africa), transitions to adulthood, young caregiving, aspirations and family change.

Sarah de Leeuw is Associate Professor at the University of Northern British Columbia, Canada, and Research Director of the Health Arts Research Centre. She is an award-winning researcher and creative writer whose work focuses broadly on marginalised peoples and geographies. Her research sits at the crossroads of social-cultural geography, health-humanities, social determinants of health, and decolonising Indigenous method/ologies. She is interested in why some peoples and communities have 'better' lives than others, and why other peoples and communities live with burdens of poverty, isolation, violence, discrimination, racism, sexism or poor health.

Tom Disney is a Research Fellow in Carceral Geography at the University of Birmingham, UK. His research interests lie at the intersection of Carceral and Children's Geographies considering spaces of youth care, control and confinement. He recently completed his doctoral research which provided an ethnography of the spaces of care for orphan children in the Russian Federation. He is currently contributing to the ESRC-funded project 'Breaking the Cycle': Prison Visitation and Recidivism in the UK.

Ruth Evans is an Associate Professor in Human Geography at the University of Reading, UK. Her research focuses on family geographies and care ethics, particularly in relation to bereavement, chronic illness and migration; and on gender and generational inequalities in land access and inheritance. She is currently leading a research project, *Death in the Family in Urban Senegal: Bereavement, Care and Family Relations* (funded by The Leverhulme Trust). She recently published a book focusing on children caring for parents living with HIV in global perspective (with Saul Becker, *The Policy Press*). Her articles have discussed sibling caringscapes *(Geoforum)* and care ethics in families affected by HIV *(Emotions, Space & Society)* and in interactions between professionals and disabled forced migrants (*British Journal of Social Work*). She has also recently published articles on women's and young people's land rights and inheritance in Senegal *(Gender, Place and Culture)* and Ghana (*Geoforum*). Ruth also has an interest in emotions in research, participatory methodologies, dissemination and 'impact'. She is co-editing the *Methodological Approaches* volume of the Springer Major Reference work, *Geographies of Children and Young People* (with Louise Holt, Editor-in-Chief: Tracey Skelton). She is an editorial board member of the journals *Gender, Place & Culture* and *Social & Cultural Geography*.

Karen R. Fisher is a disability policy researcher at the Social Policy Research Centre, UNSW Australia. Her research interests are the organisation of social services in Australia and China, including disability and mental health services and community care; inclusive research and evaluation methodology; and social policy process.

Anne Graham is Professor of Childhood Studies and Director of the Centre for Children and Young People (CCYP) at Southern Cross University, Australia. Anne's background is in education and sociology. Her research and professional interests include children and young people's rights (especially concerning their participation) and their social and emotional well-being; ethical issues in researching with children and young people; and approaches to professional learning of those working with them.

Tracy Hayes teaches primarily on the Working with Children, Young People and Families undergraduate programme at the University of Cumbria, UK, drawing extensively on her experiences as a professional youth and community development worker. She is also a doctoral researcher in the Department of Science, Natural Resources and Outdoor Studies researching the relationship that young people have with the natural environment. She takes a playfully narrative approach to her work, embracing creative methodologies, particularly reflexive auto-ethnography, creative non-fiction and story-telling. She enjoys being, and playing, outside.

Malcolm Hill is Emeritus Professor of Social Work at the University of Strathclyde, UK. Formerly he was the Director of the Centre for the Child & Society at the University of Glasgow, UK. He has written widely on the care of children and young people in biological, foster and adoptive families.

Kathrin Hörschelmann is a Researcher based in the Leibniz-Institut für Länderkunde, Germany. Kathrin is a Geographer with research interests in youth identity, globalisation, post-socialist transitions, masculinities and critical geopolitics. She co-edited *Children, Youth and the City* (Routledge, 2012) and is author of a forthcoming book titled *Children, Securitization, War and Peace* for Routledge's *Spaces of Childhood and Youth* series. Other publications include papers in international journals such as *Transactions of the Institute of British Geographers*, *Political Geography* and *Social & Cultural Geography*.

John Horton is Associate Professor in Human Geography and Associate Director of the Centre for Children and Youth (CCY) at the University of Northampton, UK. He is currently Editor of two international academic journals: *Children's Geographies*, and *Social & Cultural Geography*. He is co-author of the book

Cultural Geographies (2013) and Series Editor of Routledge's major new book series on Spaces of Childhood and Youth. He has been co-investigator on more than thirty funded multidisciplinary research projects with children and young people in the UK.

Gill Hughes is a Lecturer and doctoral researcher at the University of Hull, UK. She developed the University's BA Hons Social and Community Studies programme and is a member of the BA Hons Youth Work and Community Development team. Having worked in partnership on projects to re-engage young people with education through informal learning practices, Gill became interested in young people's experiences of (dis)connection with education. This also encompasses her interest in the increasing erosion of care in welfare services under neoliberalism, and became central to her doctoral research. She is currently researching around reimagining alternative ways of working with young people and communities and more generally, community development/community participation/youth work. Prior to roles at the University of Hull, Gill worked in the community mental health sector, offering creative and supportive voluntary sector provision to meet the needs of people with mental health issues.

Ann Pairman is a doctoral student in Education at Victoria University of Wellington, New Zealand. Ann has more than 35 years' experience as an early childhood teacher, a professional development advisor, and as a policy analyst for New Zealand's Ministry of Education. Her current research focuses on relationships between built environments and children's lived experiences in early childhood education (ECE) settings. Ann's past publications relate to ECE built environments, group size and quality in ECE, and government regulation.

Rebecca C. Pratchett is a doctoral researcher in the College of Human and Health and Sciences at Swansea University, UK. Her research focuses on 'kinship care', or 'family and friends care', in the UK with a particular focus on South Wales. She is currently working in close collaboration with the City and County of Swansea and Carmarthenshire County Council to examine the relative wellbeing and outcomes of children and young people in kinship care in South Wales, and inform the development of strategies to maximise placement success.

Michelle Pyer is a Senior Researcher in the Institute for Health and Wellbeing at University of Northampton, UK. She has been co-investigator on a large number of research projects exploring the needs of carers and families of disabled people in diverse contexts. Her doctoral research considered the care, play, leisure and wellbeing experiences of teenage wheelchair users. Michelle also has wide-ranging experience of working in/running playwork settings and as a support worker for children with additional needs in home and educational settings.

Paul Rees is Associate Professor of Public Health, Policy and Social Sciences at Swansea University, UK. He is Co-Director of the University's Centre for Children and Young People, and has extensive experience in the fields of clinical paediatric neuropsychology and educational psychology and child psychology. His most recent research has explored the health literacy and resilience of children and young people who are fostered or 'looked after' in the UK.

Sally Robinson is a Senior Research Fellow at the Centre for Children and Young People (CCYP), Southern Cross University, Australia, where she does participatory and inclusive research with young people with disability and those who support them. Her interests centre on the use of creative methods to engage people with cognitive disability in research to include their perspectives on key issues central to their safety, rights and wellbeing.

Anna Schliehe is a postdoctoral researcher in Criminology at the University of Cambridge, UK and did her PhD project in Human Geography at the University of Glasgow, UK. Her research on the Scottish criminal justice system and its responses to young women in particular is informed by both carceral geography and criminological scholarship. Anna is interested in understanding the nature and experience of closed spaces, connecting empirical to conceptually challenging research on concepts of care and control.

Rowena M. Taylor maintains a part-time role at the Institute of Education, Massey University, Palmerston North, New Zealand where she was formerly a Senior Lecturer. She draws on many years of experience as a high school Geography and Social Studies teacher and as a teacher educator to inform her research in social sciences curriculum, pedagogy and assessment. Her recent research centres on citizenship education and subject-specific literacy practices in the senior social sciences. Her research features in a wide range of New Zealand journals as well as in edited texts on social sciences pedagogy.

Karen Wells is Reader in International Development and Childhood Studies, and Programme Director for Children, Youth and International Development at Birkbeck, University of London, UK. She is the author of *Childhood in a Global Perspective* (Polity, 2009) and has published widely on children and visual culture. Her research focuses on how global processes and structures interact with practices of childhood in local contexts and how this dialectic between the global and the local impacts on children's lives and on practices of childhood. She is currently researching the life history narratives of children who have been fostered into the UK from overseas (privately fostered children) in collaboration with the charity, Children and Families Across Borders (CFAB).

Samantha Wilkinson is a postdoctoral Research Fellow in the Institute of Mental Health, School of Sociology and Social Policy, University of Nottingham, UK. She is currently working on a mixed-methods qualitative research project aiming to broaden understanding of good home care for people with dementia. Prior to this, she was a Human Geography PhD student at The University of Manchester, UK, having previously completed a BA in Human Geography and an MSc in Environmental Governance at the same institution. She has presented findings from her thesis on 'young people, alcohol and urban life' at a number of national and international conferences. She has published a range of journal articles and book chapters on the themes of young people friendly mobile methods; young people's drinking spaces; and young people's alcohol-related im/mobilities.

Bronwyn E. Wood is a Senior Lecturer in the Faculty of Education, Victoria University of Wellington, New Zealand. Her research interests lie at the intersection of Sociology, Geography and Education and centre on issues relating to youth participation, citizenship and education. Her recent research focuses on young people's social action in schools, restorative practices in cities and education and global education policy. Her work has been published in diverse journals including *Emotion, Space and Society*, *Children's Geographies*, *Political Geography* and the *British Journal of Sociology of Education*.

Acknowledgements

Many chapters in this book were first presented at a series of sessions on *Children, Young People and Spaces/Practices of Care* which were organised by the editors at the Annual Conference of the Royal Geographical Society / Institute of British Geographers (RGS-IBG). We thank the RGS-IBG's Geographies of Children, Youth and Families Research Group and Geographies of Health Research Group for jointly sponsoring the sessions. We would also like to thank all those who attended and participated in the sessions for such a rich, thought-provoking series of discussions.

We are really grateful to Faye Leerink and Priscilla Corbett at Routledge/Taylor & Francis who have been so supportive, helpful and efficient in overseeing the development and completion of this book. Thanks also go to two peer reviewers who provided generous, careful, detailed and insightful comments on first drafts of all of this book's chapters.

Thanks are also due to the authors of each chapter for their care and timeliness in preparing and revising manuscripts. We are particularly grateful to Sophie Bowlby and Kathrin Hörschelmann who have found time to write such thoughtful discussant pieces reflecting on the edited collection as a whole.

We are grateful, as ever, for collaborators including Faith Tucker, Peter Kraftl, Sophie Hadfield-Hill, Alison Ward and Helen Lomax for providing caring intellectual spaces which supported the development of this book and much else besides. Thanks are also due to our past and present doctoral students. We also thank colleagues in the School of Social Sciences, School of Health, and Institute for Health and Wellbeing at the University of Northampton, UK for their support and understanding during the production of this book.

As indicated in the main text:

- In Chapter 1, longer quotations from Lawson (2007) and Bowlby (2011) are reproduced with permission from Taylor and Francis Rightslink, under license numbers 501208192 and 3993550054094 (both November 2016) respectively. We are also grateful to Sophie Bowlby for permission to use excerpts from her work on caringscapes.

- Chapter 4, by Sarah de Leeuw, was previously published as de Leeuw, S. (2014) State of care: the ontologies of child welfare in British Columbia. *Cultural Geographies* 21, pp.59–78. We thank Sage Publications for permission to republish this manuscript under license number 3854450276476 (April 2016).
- In Chapter 10, by Ruth Evans, Figure 10.1 has previously appeared in Evans, R. (2011) 'We are managing our own lives…': life transitions and care in sibling-headed households affected by AIDS in Tanzania and Uganda. *Area* 43, pp.384–96. We thank John Wiley and Sons for permission to republish this image under license number 3847671247604 (April 2016).

1 Introduction

Children, young people and 'care'

John Horton and Michelle Pyer

Preface: three articulations of childhood, youth and 'care'

> Warm, sensitive, and responsive care is the foundation of sociomoral behaviour.... Parenting and communal practices such as breastfeeding, touch, alloparenting, and play ... are associated with important physiological and psychological outcomes that significantly raise the likelihood of a child interacting with others in prosocial ways. Most important, a community of individuals who receive the kind of parenting that is associated with sociomoral flourishing is more likely to create a community in which joy and well-being are fostered by and for all.
>
> (Gleason and Narvaez 2014, p.335, 345)

> Today, President Obama outlined his plan to make affordable, quality child care available to every working and middle-class family with young children. His plan includes: making a landmark investment ... that helps every eligible family with young children afford high-quality child care ... [and] creating a new innovation fund to help states design programs that better serve families that face ... challenges in finding quality care.
>
> (White House 2015, unpaginated)

> **Young people don't care about politics ... or do they?**
>
> With the EU referendum just months away, campaigners from both sides of the debate are trying to energise and engage with 18 to 24-year-olds so that they vote.... Their turnout in political elections is usually lower than any other age group.... Research suggests a significant proportion are not interested in politics and do not turn out to vote.
>
> (BBC 2016, unpaginated)

Starting points: contextualising 'care'

In so many contexts, and in all kinds of ways, when we think about childhood and youth we are always already thinking in terms of 'care'. The very notions of childhood and youth are intimately, normatively and fundamentally connected to contemporary assumptions, discourses and spaces of care, caring and caregiving. Thus, as suggested by our preface, we might find ourselves habitually working to provide safe, nurturing, happy, responsive, sensitive environments for children or young people in our care, or debating the merits or limitations of policy interventions relating to child-care or youth support, or critically reflecting upon the extent to which children and young people (do not) care about democratic processes and political issues. As the first prefatory example articulates, the provision of care is widely figured as both *the* primary responsibility of parents, carers and practitioners who work with children and young people, and *the* primary factor in shaping children and young people's development, education, socialisation, wellbeing and contentment. While specific practices and spaces of care may be socially and historically contingent (as evidenced by diverse anthropological and historical examples of familial, domestic, cultural or communitarian behaviours in relation to care) the expectation that children and young people should be 'cared-for', 'nurtured', 'supported' and allowed to 'flourish' by adults may be instinctual, and is certainly so widely recurrent as to appear essential and fundamental. This normative, discursive responsibilisation of adult *care-for* children and young people is widely reproduced within extensive frameworks of state policy interventions, legal/statutory programmes and institutional spaces devoted to care-provision. In diverse contexts, familial and domestic caring practices are increasingly shaped, regulated and mediated by local, regional, national and supranational policy, legal and institutional infrastructures. And, as in our second prefatory example, these are often explicitly named as spaces of, or plans for, 'care': as in the global preponderance of policies relating to 'child-care' or 'social care', or services provided for young people 'in-care' over the last century. These complex, shifting and often-precarious geographies of formalised care, welfare and service provision are instrumental (for better or worse) in constituting the everyday lives of many children, young people and families. However, as the third example on the preface suggests, children and young people themselves are rarely figured as key actors in the provision of care. In many contexts, an overwhelming presumption that children and young people are to be cared-for has effectively marginalised their agencies and responsibilities as carers, or in relation to practices and spaces of care. More widely, young people (perhaps particularly teenagers) are too often represented as *uncaring* and apathetic in relation to all manner of political, social and community issues.

This edited collection draws together new research exploring these kinds of interconnections between childhood, youth and care. Although chapters are predominantly written by researchers working in the disciplinary tradition of Human Geography – representing the significant array of geographical work on children, young people and families (see Kraftl et al. 2014, Skelton et al. 2016, and the international journal *Children's Geographies*) – they engage with a remarkably diverse range of conceptual and empirical work on the topic of 'care'. In the following chapters, the term 'care' surfaces in different ways in diverse research contexts and via parallel theoretical or methodological traditions. Indeed, a key point of this book is to demonstrate and juxtapose some of the complex, multiple ways in which care happens, matters and is articulated in diverse scholarly contexts, and in different children and young people's lives. It is therefore hard, and arguably problematic, to settle on a neat, singular definition of 'care' (see Bowlby et al., 2010, pp. 39–41, Milligan and Wiles 2010, pp.737–8). However, as a minimal and inclusive point of departure we understand care, in the geographer David Conradson's (2003a, p. 508) terms, as "the proactive interest of one person in the well-being of another". As Sophie Bowlby (2011, p. 618) notes, in practice this definition encompasses a wide range of tasks, relationships, situations and modes of "intimacy, care and support that people use to get through the vicissitudes of life, both major and minor". Many authors have attempted to map the multiple ways in which care exists and matters in different situations, relationships and practices (see Barnes 1997, 2012, Noddings 2013, Rogers and Weller 2013). For example, the feminist political scientist Joan Tronto's (1993) fourfold typology of phases and capacities of care has been widely influential. As Table 1.1 summarises, Tronto (1989, 1993, 2013, also Fisher and Tronto 1990) argues that care can entail at least four different kinds of activity, and at least four distinctive, corresponding kinds of moral-ethical disposition (note that although these fundamental activities are termed 'phases', Tronto recognises that they very often overlap, repeat, intersect and occur in complex nonlinear sequences).

As we outline in the following section, this kind of typology has been the basis for a series of important critical theorisations of the politics and ethics of care. In particular, a great deal of social scientific work begins from a distinction between what are typically termed *caring-about* (corresponding to the first phase of Tronto's typology) vis-à-vis *caring-for* (approximately corresponding to the other phases). As Milligan and Wiles (2010) explain, *caring-for* "implies a specific subject as

Table 1.1 Tronto's (1993) typology of phases and capacities of care

Phases of care	*Corresponding capacities of care*
Caring about: an initial noticing and recognition of the need for some form of care, and a preliminary assessment of how this need could – or *should* – be met	*Attentiveness*: noticing the need for care; permitting oneself to "suspend one's own goals, ambitions, plans of life, and concerns, in order to recognise and be attentive to others" (p.128)
Taking care of: assuming a degree of responsibility for addressing, responding to, or setting in motion a sequence of actions in response to a need for care	*Taking responsibility*: accepting and taking up responsibility for some form of care, particularly beyond one's minimal pre-existing everyday obligations
Care giving: practically acting in response to care needs, *doing* the immediate, practical, 'hands-on' provision of care, as required	*Competence*: cultivating skills and habits to ensure that care needs are met adequately, effectively, appropriately and efficiently
Care receiving: some form or degree of reciprocity and acknowledgement from care recipient(s), evidencing that care needs are being met	*Responsiveness of care recipient*: providing evidence or acknowledgement that care needs have been appropriately attended-to, taken-responsibility-for, and competently addressed

the focus of caring" whereas *caring-about* comprises "a more general form of commitment that refers to less concrete objects". Thus while caring-about is theorised as a somewhat generalised disposition toward, or feeling of, *being caring*, the more specific work of caring-for may encompass diverse

> activities undertaken by formal paid workers or informal, unpaid workers such as family, friends and volunteers. It exceeds the norms of reciprocity commonly practiced between adults.... At its most fundamental, caring *for* is about the personal, the performance of proximate and personal care tasks but it can include other everyday tasks such as childminding, pet care or household tasks. At a distance it can involve arranging and monitoring paid and professional care. (Milligan and Wiles 2010, pp.740–1)

Or, moreover, as Bowlby (2011) elaborates,

> [care] can involve 'caring for' – that is tasks of care – as well as 'caring about', which refers to emotional investment in another person's problems and concerns. Clearly the two are related but 'caring about' does not necessarily lead to 'caring for', while 'caring for' can occur without 'caring about'. However, the experience of 'caring for' often leads to 'caring about' someone. Care is part of many different kinds of social relationship – for example, care can be paid or unpaid, formal or informal, and may result from social obligation, lover, fear or because of a need for income. (p.606)

The sociologist Carol Thomas (1993) further highlights the complexity of the term 'care' by identifying seven dimensions which characterise any caring practice. The seven dimensions are listed in Table 1.2 alongside some prompts for reflection: you may find it productive to use these prompts to think about how any particular act of care (perhaps one from your own work, or from the following chapters) is distinctly patterned in relation to each of the seven dimensions.

As Bowlby et al. (2010) note, the dimensions listed in Table 1.2 can combine in different ways so that different events of care are patterned and differentiated as a result of the distinctive, contingent constituent features of carers, care-recipients, care practices and social-economic and spatial-institutional contexts. Thus, for example, these dimensions can combine to constitute types of care as diverse as "formal elderly care in America provided by a private firm; informal care exchanged among family members in Africa, or 'free' childcare provided by a state-run nursery in Europe", and ranging across all manner of "domains in which care takes place (such as the family, workplace, hospital) as well as ... many types of relations (including intimate partner, relatives, friends, work colleagues)" (Bowlby et al. 2010, p.40). Even more complexity and multidimensionality is suggested by a wide range of studies which identify diverse caregiving 'styles' among individuals, care professionals, or organisations. Corcoran's (2011) typology of family caregiving styles in relation to elderly relatives in one particular

Table 1.2 Thomas's (1993) seven dimensions of care

Dimensions of care	*Our prompts for reflection*
Carer	Who is the carer? What are their characteristics, personality and positionality? What is their location, age, gender, religion, ethnicity, social class, profession, 'dis/ability' or social-cultural, educational and geographic background? How might these characteristics matter in the provision of care?
Care-recipient	Who is the care-recipient? What are their characteristics, personality and positionality? What is their location, age, gender, religion, ethnicity, social class, profession, 'dis/ability' or social-cultural, educational and geographic background? How might these characteristics matter in the receipt of care?
Relationship between carer and care-recipient	What is the nature of the relationship between carer and care-recipient? How might different relationships (e.g. familial, parental, sibling, marital, romantic, friendship, colleague, acquaintance, formal, professional, contractual or institutional) constitute different forms or experiences of care?
Nature of care	What form is care taking in practice? Which phases or capacities of care (see Table 1.1) are most evident? How might the style of care (see Table 1.3) be characterised?
Social domain	How is the relationship between carer and care-recipient connected to other individuals or social groups? How do these connections support or hinder the care process? How is care contextualised and affected by local, national or global social-demographic factors? How do social-cultural norms, assumptions, ideals or inequalities matter?
Economic character	How is care constrained or afforded by economic conditions? How do economic inequalities shape the capacities of carers and the experiences of care-recipients?
Institutional setting	More broadly, where does care take place, and how does this matter? How do spatial and environmental characteristics of spaces of care impact upon carers and care-recipients?

study in North America (see Table 1.3) provides just one illustration of ways in which diverse caring dispositions, habits and preferences may be constituted by individuals' personality, positionality, upbringing, experiences, resources, behavioural styles, social capital and/or social-environmental factors.

Having contextualised the topic and practice of care, the following sections of this chapter outline why care has come to be a core concern of much multidisciplinary research over the last two decades. We then develop an argument that recent scholarly work on care poses some particularly important challenges and questions for researchers and practitioners engaging with children, young people and families in diverse contexts. In the chapter's final section we outline how the

Table 1.3 Four styles of dementia-related family caregiving in Washington, DC (Corcoran 2011)

Facilitating: "characterised by a focus on the emotional health of the care-recipient.... A facilitating style involves promoting one or more of the following: (a) an image of the care-recipient as a productive adult and loving spouse, (b) the care-recipient's ability to enjoy meaningful activities.... A hallmark of a facilitating style is a preference for one-on-one, cooperative activities involving working together (e.g., hobbies, cooking, or gardening)... supporting the emotional health of the care-recipient through reassurance and by avoiding upsetting experiences". (p. 467)	*Balancing*: "enacted through use of environmental controls (alarms, intercoms, and baby monitors) to supervise the care-recipient, restrict movement, and avoid or eliminate problems.... Environmental controls may also include modifying items, such as removing knobs on appliances, installing locks, camouflaging doors with curtains or wallpaper, or providing interesting distractions... that require only distant supervision by the caregiver. Favoured activities are typically repetitive and highly familiar, including household chores such as sweeping, polishing, and folding.... While the care-recipient is occupied, caregivers are free to work or relax nearby". (p.468)
Directing: "defined by an overall focus on the physical health of the care-recipient, including his or her nutrition, medical routines, and hygiene... [and] a predominant tendency toward maintaining order and certain standards of behaviour.... These standards include good nutrition, exercise, adherence to medical regime, and staying independent. Strategies to enforce these behavioural standards vary, although for the most part they are communicated verbally and reinforced whenever the care-recipient's behaviour seems dangerous, unhealthy, or needy". (pp.469–70)	*Advocating*: "distinctive for being vigilant about the care-recipient's well-being and advocating when problems are suspected.... Advocating style caregivers monitor their care-recipient's need for help and are careful not to provide more assistance or support than is needed at a particular point in time.... An advocating style often requires that caregivers interact with the care-recipient's formal providers. This may be done to get information, make requests, or assure that staff knows someone is 'keeping an eye on things'". (pp.468–9)

authors represented in this collection contribute to this context via a range of new theoretical discussions, research findings and critical reflections on the question of children, young people and care.

So, why care?

The chapters collected in this volume have emerged from, and extend, a large body of multidisciplinary research on diverse aspects of care over the last two decades. For example, one can trace sustained turns to address questions of care in academic disciplines as diverse as Anthropology, Childhood Studies, Early Years Practice, Education, Gender Studies, Gerontology, Geography, History, Medicine, Nursing, Occupational Health, Organisational Studies, Philosophy, Playwork, Political Science, Psychology, Social Care, Social Policy, Social Work, Sociology and Youthwork, as evidenced by a substantial array of interdisciplinary edited collections (Brechin et al. 1999, Rossi 2001, Dermott and Seymour 2011,

Engster and Hamington 2015), literature reviews (Parr 2003, Atkinson et al. 2011, Milligan and Wiles 2010) and special issues of scholarly journals (Conradson 2003b, Staeheli and Brown 2003, Raghuram et al. 2009, McEwan and Goodman 2010, Zhang and Jean Yeung 2012). This body of work has produced a substantial range of important, rich, thought-provoking material which has challenged researchers and practitioners to *care more about care*. Here, we highlight three particularly important challenges which recur across this multidisciplinary context.

First, multidisciplinary work on care has made plain the significance of caring practices in/for individuals' everyday lives, and also in the ongoing constitution of relationships, homes, families, communities, solidarities, friendships, workplaces, institutions, localities, activisms, civilities, cosmopolitanisms, citizenships, public spaces, and all manner of social-cultural collectives. As we have already suggested, everyday caring practices may take the form of intimately co-present, proximate, embodied, microgeographical encounters, perhaps encompassing,

> help with personal and intimate activities such as bathing, toileting, or dressing, and activities such as preparing meals, shopping, arranging travel, or manage finances …, everyday acts of practical and emotional support. (Wiles 2011, p.574; see also Milligan 2000)

Alternatively, everyday caring practices may be constituted via more spatially complex or translocal practices (Brickell and Datta 2011), or via particular ethical dispositions towards 'distant others' (Silk 1998, 2004) or encounters with 'other' people, places or co-presences (Amin 2002, Barnett 2005, Laurier and Philo 2006, Bell 2007). Research on diverse forms of everyday caring practice has evidenced how such activities matter to individuals (e.g. in terms of emotional significance, preoccupation of time and energy), constitute manifold social-relational connections, and are important foci of multiple forms of paid and unpaid work. However, despite their substantial personal, psychological, socio-cultural, economic, societal and political importance there is considerable evidence of the extent to which everyday caring practices have been significantly, effectively systematically, marginalised, undervalued and concealed within most societal formations (Lawson 2007). A key accomplishment of feminist scholarship on care has been to politicise everyday caring practices to contest this inequity. Thus, for Tronto (1993),

> care as a political concept requires that we recognise how care – especially the question, who cares for whom? – marks relations of power in our society and marks the intersections of gender, race and class with care-giving. (pp. 168–9)

It remains the case that answers to the question 'who cares for whom?' reveal overwhelmingly, inequitably gendered, classed and raced divisions of labour: in so many contexts, everyday caring practices are disproportionately done by women and marginalised social groups, and many kinds of caring practices done by women and marginalised social groups are disproportionately undervalued, low-paid or unpaid. In the UK for example, domestic care for children and relatives, employment in jobs which entail hands-on care-provision (e.g. nursing), participation in voluntary work which involves caring practices, and 'extracurricular' taking-on of caring roles (e.g. within educational institutions) all remain significantly, and often overwhelmingly, gendered and politically-culturally-economically undervalued (Silk 1998, ONS 2013, HSCIC 2014, Skills for Care 2015). Thus feminist theorists of care have been important in articulating how, overwhelmingly,

> care work is women's work. Paid and unpaid, located at home, in voluntary organisations or in the labour force, the overwhelming majority of care is provided by women. It is often invisible, usually accorded little value and only sometimes recognised as skilled.
>
> (Armstrong and Armstrong 2001, p.1)

Drawing on longstanding feminist critiques of the (non)place of unpaid domestic and maternal labour in contemporary, international economic systems (see Boserup 1970, Young 1981, Glazer 1984, Nicholson 1988, Gardiner 1997) many authors, such as England (2010), historicise the marginalisation of everyday care – and separation of '"work"-production-public' from 'care-reproduction-private' practices as follows:

> [The] gendering of work has its ideological and historical roots in the social and spatial separation of waged work from social reproduction. 'Work' became constituted as 'economically productive' waged-labour that took place outside the home. Housekeeping, caring for family members and other domestic activities became non-marketised 'labours of love' primarily associated with the private sphere of family and home, and assumed to be primarily the responsibility of wives and mothers. (p.133)

A second major contribution of recent multidisciplinary research on care has been to highlight and explore the complex spatialities of care. For example, an important lineage of work has explored the social-political construction of, and care experiences within, designated spaces and institutions of care. There are rich and affecting seams of research on spaces such as asylums (Philo 1997, 2004), hospitals (Allen 2001, Rowland 2014), hospices (Vivat 2002, Brown 2003), therapeutic landscapes (Williams 2007), mental health services (Kearns and Joseph 2000, Parr 2000, 2008), food banks (Lambie-Mumford and Green 2015), homeless shelters (Williams 1996, Brinegar 2003, Cloke et al. 2008, Daya and Wilkins 2013), and community outreach/care services such as those delivered via faith-based or 'post-secular' organisations (Conradson 2003a, Cloke and Beaumont 2013). As well as vividly evoking daily life in these kids of spaces, this line of research has been significant in evidencing how spaces of care – and care-work therein – continue to undergo transformation as a result of processes of professionalisation, commodification, neoliberalisation, regulation and economic restructuring (Williams 2002, Smith 2005, Green and Lawson 2011, Hall 2011). However, as Hall (2011) notes, social scientific research on care has also increasingly shifted its focus beyond these formally designated sites of care to develop an expanded sense of the manifold spaces where care happens (e.g. homes (see Dyck et al.

2005, England 2010), or informal or deinstitutionalised sites of community-based care), and also the complex spatialities of care itself. For example, this latter shift might be detected in recent studies which carefully detail complex geographies such as global migrations of care-workers (Dyer et al. 2008, Huang et al. 2012), translocal caregiving within diasporic communities (Datta et al. 2010), or caring (and indeed *uncaring* and exclusionary) gestures within urban public spaces (Valentine 2008, Laurier et al. 2002, Laurier and Philo 2006). A number of Human Geographers working in this context have sought to develop broader theorisations of the spatialities of care. For example, Milligan and Wiles (2010, p.740) use the term *landscapes of care* to think through the "complex and organisational spatialities that emerge from and through the relationships of care". They describe the concept of *landscapes of care* as part and parcel of attempts to,

> articulate care through the differing, and sometimes surprising, social spaces that enable caring interactions …; as individual space-time trajectories through varied social landscapes of care, care-giving roles, employment and social policies, and gendered and generational expectations of care and work …; through the entanglements of inclusion and exclusion, dependency and independency that can manifest within and across formal and informal space of care …; and through the emotional landscapes that underpin care and care interactions.
>
> (Milligan and Wiles 2010, p.738)

They note that these landscapes typically traverse diverse institutional, domestic, familial, community, public, voluntary, private, interpersonal and international geographies. Similarly, Sophie Bowlby and Linda McKie use the linked concepts of *caringscapes* and *carescapes* to prompt reflection upon the complex spatialities of any caring practice. They describe *caringscapes* as the

> shifting and changing multidimensional terrain that comprises people vision of caring responsibilities and obligations: routes that are influenced by everyday scheduling, combining caring work with paid work and the paid work of carers.… People create routes through 'caringscapes', which change and evolve as they move through the lifecourse.
>
> (McKie and Bowlby 2004, p.2)

Since the concept of *caringscapes* is a touchstone of many chapters in this book, Box 1.1 presents a fuller introduction to this idea.

Box 1.1 Navigating caringscapes

The followig quotations provide a fuller introduction to the concept of caringscapes, which is touchstone of many sections of this book:

> At different times in people's lives working and caring mutually influence one another through the formation of families, the bearing of children, the changing

(continued)

> needs of kin or friends for assistance in illness and old age, and participation in education, training and employment. The interaction between the private – the space in which caring work for the very young is located – and the public – where paid work and caring for older children is located – involves a complex system of physical manoeuvres most obviously linked by processes of 'getting to work on time' and 'leaving work in time to pick the kids up' or 'making time' to check up on the frail elderly parent. This process of thinking and dealing with different places and journeys throughout the day and week can be thought of as a system of spatial reference. These everyday realities are informed by past experiences and future anticipation of caring. Caringscapes can be thought of as shifting and changing multi-dimensional terrain that comprises people's vision of caring possibilities and obligations: routes that are influenced by everyday scheduling, combining caring work with paid work and the paid work of carers…. People create routes through 'caringscapes', which change and evolve as they move through the lifecourse.
> (McKie and Bowlby 2004, p.2)
>
> As a person moves through a terrain, s/he can look ahead and plan a direction and a route across this varied landscape…. Some travellers do not plan their route but follow the main path without thinking or follow a path marked out for them by others. Many plan some sections of their journey but leave others to chance.
> (Bowlby et al. 2010, p.6)
>
> A person does not travel across this landscape alone: their fellow travellers will change over time and will affect people's choice of routes…. [P]arts of the terrain might not be clearly visible to the traveller and… like the landscapes in Alice in Wonderland, the terrain might change over time as the social and economic environment alters. The metaphor [of caringscape] draws attention to present and future time but can also be used to illustrate the significance of past time to current decisions through the influence of memories and habits of travel and the cumulative impact of past choices of route on current possibilities.
> (Bowlby 2012, p.2110)

Bowlby (2012) also deploys the term *carescape* to denote one particular set of features of the *caringscape* terrain. Thus *carescape* specifically refers to,

> the resource and service context shaping the 'caringscape terrain'. Thus we might talk about the 'carescape' of a local government area, a nation state or a particular organisation…. For example, in a recent project … we have begun to flesh out the notion of carescape by showing how the 'organisation carescape' – the care policies and practices of employing organisations – affect the caringscapes of individual employees within and beyond the workplace.
> (Bowlby 2011, pp.2112–3)

Taking these concepts together, Bowlby et al. (2010) present a list of propositions about caringscapes which are designed as a framework to assist researchers, practitioners and policy-makers in reflecting upon caring practices. Their eight propositions are reproduced in Table 1.4.

Table 1.4 Bowlby's (2012, p.2112) eight propositions about caringscapes

Number	*Proposition*
1	"Informal caring is a social/relational activity embedded in notions of obligation and reciprocity"
2	"Informal caring is an ethical activity involving notions of normative behaviour and desirable or 'appropriate' social relationships"
3	"Informal caring often involves relationships of unequal power – especially inequalities of class, gender, race and (dis)ability"
4	"Informal caring is an embodied activity in which copresence, bodily separation, and mediated forms of communication affect the caring encounter"
5	"The performance of informal caring necessarily involves processes that connect across time-space"
6	"The performance of informal caring involves a person using resources – people, ideas, material objects, as well as time and space to care – accessed through particular places and spaces at particular times"
7	"The performance of informal caring involves a person reflecting on or enacting aspects of past informal caring relationships and/or anticipating future caring relationships"
8	"The performance of informal caring involves a person reflecting on and/or anticipating how informal caring relationships link to other actual or anticipated aspects of their lives"

As Bowlby et al.'s propositions about caringscapes as relational, ethical, political and critically-reflective suggest, the topic of care has increasingly been a prompt for wide-ranging and challenging political-ethical reflection among social scientific researchers. Indeed, a third major contribution of multidisciplinary work on care has been to focus discussion on the present and potential ethics of care. Thus, as Atkinson et al. (2011) note, the concept of care has been an important point of mobilisation for politically-charged feminist, political and communitarian critiques of contemporary societies and academic practices. In particular, theorisations of an *ethic of care* have been powerfully influential. Feminist scholars including Tronto (1993, 2013; see also Table 1.1) have long argued that normative contemporary social and political structures could be reimagined and recast if greater value were given to caring relationships. As outlined in the preceding section, Tronto (1993) focuses attention on the practical and ethical underpinnings of caring practices. It is also noteworthy that her vision of an ethic of care is predicated upon a wide-ranging conceptualisation of caring responsibilities, such that,

> caring [should] be viewed as a species activity that includes everything we do to maintain, continue and repair our 'world' so that we can live in it as well

> as possible. That world includes our bodies, our selves and our environment, all of which we seek to interweave in a complex, life-sustaining web. (p.103)

This vision of an ethic of care both challenges individuals to foster an everyday 'habit of care' in relation to the bodies, selves, co-presences and environments encountered in daily life (Tronto 1993, p.127), and demands a broader recasting of assumptions about citizenship which,

> requires that citizens take seriously the responsibility for 'caring with' each other.... To do so requires a change in the values of citizens. It requires that citizens care enough about caring – both in their own lives and in the lives of their fellow citizens – to accept that they bear the political burden of caring for the future. (Tronto 2013, p.xii)

This ethic of care is therefore powerful in envisioning social-political arrangements focused around personal and communitarian relationships of care, in stark contrast to many normative prevailing social-political inequities (Sevenhuijsen 1998, McDowell 2004, Smith 2005, Held 2006, Lawson 2007). For Milligan and Wiles (2010), thinking with the concept of *ethic of care*,

> could be a framework not just for understanding who gives care, where and why ... but also for understanding how an approach informed by care might enlightened our entire way of collective and individual being. (p.743)

The implications of this way of thinking – which Milligan and Wiles characterise as fostering more '*care-ful*' or 'passionate' modes of work – are potentially wide-ranging. Lawson (2007) describes the visionary, hopeful criticality of an *ethic of care* in the following, affecting terms:

> What do I mean by care ethics? ... Care ethics questions (neo)liberal principles of individualism, egalitarianism, universalism, and of society organised exclusively around principles of efficiency, competition and a 'right' price for everything. Under neoliberal principles, care is a private affair, occurring in homes and families. In the privatization of care, we construct certain sorts of people as in need of care – the infirm, the young/elderly, the dependent, the flawed – ignoring the fact that we, all of us, give and need care. By contrast, a feminist ethic of care begins from the centrality of care work and care relations to our lives and societies. Care ethics begins with a social ontology of connection: foregrounding social relationships of mutuality and trust (rather than dependence). Care ethics understands all social relations as contextual, partial, attentive, responsive, and responsible.... We see care not as a separate kind of relation, but as endemic to (potentially) all social relations that matter.... [C]are ethics is concerned with structuring relationships in ways that enhance mutuality and well-being. Care ethics also demands attention to

> emotions and affective relations (of love, concern, and connection) because of the complex ways in which power is embedded within them.... But in sum, what is most important is that care ethics suggests different ways of theorizing politics.... [W]e are not the autonomous individual imagined in liberal political theory, rather we are all profoundly interdependent, and yet not all equally burdened with the work of care. (p.3)

This book seeks to extend this rich seam of work on practices, scapes and political-ethical dimensions of care. Specifically, we have drawn together chapters which foreground and exemplify children and young people's diverse presences, roles and responsibilities in relation to care. In so doing, our broader aim is to bring multidisciplinary theorisations of childhood and youth into closer inter-relation with contemporary theorisations of care, caringscapes and ethics of care.

Children and young people *do care*

Although care has come to be a core concern of such a range of interdisciplinary academic research, we are struck by the way in which work in this context has, to date, rarely engaged directly with children and young people. Research *with* – rather than *about* – children and young people in relation to care seems to us to be, problematically, rather marginalised; certainly, it is relatively rare to encounter children and young people's *own* voices, experiences, issues, practices, politics and ethics represented within the very rich body of work on care, caringscapes and ethics of care (for some important exceptions see Robson 2004, Weller 2007, Bartos 2012, Wood 2013, 2016, Day and Evans 2015, Day 2016). Evans (2012) argues that this absence is part and parcel of wider normative, minority world social-cultural constructions of children and young people as 'dependent' 'care recipients', not actively engaged in 'caregiving' themselves. We would further suggest that this assumption is compounded by a range of related social-political discourses and contemporary habits of academic research, particularly: (1) still-prevalent normative binaristic cultural, media and policy discourses of children and young people as *feared* or *feared-for* (in Valentine's [1996]; also James and James 2004) oft-cited terms, as vulnerable 'angels' or threatening 'devils', which do not readily acknowledge or accommodate children and young people's caring practices and roles; (2) a still-evident tendency (outlined in the preceding section) for everyday care (including intergenerational and intragenerational caring practices involving children and young people) to be somewhat marginalised, undervalued and concealed within many societal formations and modes of scholarship; and (3) a tendency for critically valuable work analysing macro- geographical, political and economic aspects of care (e.g. processes of care- neoliberalisation, commodification and regulation previously mentioned) to direct attention away from everyday, intimate, micro-geographical spaces in which care (and perhaps particularly many forms of care involving children and young people) *happens* and *matters*.

Against this grain, this collection brings together a range of multidisciplinary and international research which directly engages with children and young people in relation to care. The collection thus foregrounds diverse, often-neglected practices, spaces and ethics of care involving children and young people. We suggest that this focus upon children, young people and care is particularly important for three reasons. First, simply, caring practices and experiences are a significant, though often overlooked, part of many children and young people's lives (Harris and Wyn 2009, Bartos 2012, Wood 2016). Children and young people are cared for and – although this seems to be less often recognised – they *do care* in all kinds of ways, and within all kinds of caring relationships, reciprocities and caring-scapes. Geographer Ruth Evans' work has been important here, both in evidencing and highlighting the existence of children and young people's caring roles and in considering how children's and young people's care intersects with complex mobilities, politics, social-cultural changes and lifecourse transitions within and between particular case study locations (R. Evans 2010, 2012, 2015, R. Evans and Becker 2009). Taking our cue from this pioneering work, the chapters collated in this book provide further evidence of children and young people's diverse experiences and practices of care. They show how care permeates children and young people's lives as they receive care, *care-for, care-about* and habitually constitute *scapes, politics* and *ethics* of care in all manner of geographical, political and social and historical contexts.

Second, it is the case that a whole range of designated, purposely-constructed spaces of care are important for many children and young people's lives. As we noted in the preceding section, a great deal of social scientific work has explored the histories and experiences of spaces such as hospitals, mental health services, homeless shelters and spaces of community outreach and care. Many chapters in this book complicate and extend this body of research, both by considering the often-underresearched presence of children and young people in these kinds of spaces, and by highlighting the existence of many other institutions and built environments specifically designed for the care of children and young people. Thus the following chapters include consideration of care practices in spaces such as early childhood centres, classrooms, orphanages, and specialist secure care units in diverse national and political contexts. In so doing, they prompt reflection upon so many other forms of specialist space constructed for the care of children and young people: in effect, they demand further research on many other spaces of care, elsewhere and otherwise (perhaps engaging with children, young people and care in settings like spaces of refuge, crisis care, emergency care, respite care, youthwork, social work, playwork, social care, medical care, palliative care, convalescence, residential and domiciliary care, play, education, therapeutic or counselling contexts). Moreover, the following chapters also call attention to the significant array of pre-emptive biopolitical and state projects which target and regulate the care of children and young people, and underpin the construction of spaces of/for care, politically, discursively, materially and affectively (B. Evans 2010, B. Evans et al. 2011, Pykett 2011, 2015, Jones et al. 2013). So much of many children and young people's lives is lived with/in designated spaces of care – and governed via policy interventions and regulatory mechanisms which shape caring practices – so the following chapters signpost important ways of researching, understanding and theorising contemporary childhood and youth.

Indeed, third, we would like to suggest that the linked concepts of *care, caringscapes* and *ethics of care* potentially offer an important new lens for (re)thinking childhood and youth in a number of ways. We wonder whether thinking with the concepts and language of 'care' might permit discussion of children and young people's lives without immediate recourse to some normative ideas and assumptions which are habitually associated with childhood and youth. For example, we suggest that concepts of care, caringscapes and ethics of care might make it possible to write and think about children and young people without necessarily falling back upon normative and often-presumptive ideals like Family, Home, Parenthood, Education, Citizenship, Participation, Child Development, and the aforementioned 'Angel'/'Devil' dualism. The following chapters seem to us to offer a series of openings to (re)think childhood and youth more care-fully: acknowledging that children and young people's lives are full of caring relationships and practices; recognising children and young people's presence in, and relationships to, designated spaces of care; critically reflecting upon ways in which complex, diverse care practices and caringscapes are often reductively labelled and understood via a series of arguably quite limited and normatively-loaded concepts (e.g. Family); and always working towards more careful, ethical, hopeful and affirmative ways of working with children and young people, in whatever capacity. As the following collection has materialised, we have increasingly developed a sense that consideration of care can unsettle many prevalent adultist ways of knowing and researching childhood and youth. Perhaps most immediately, the acknowledgement that children and young people *do care* – for each other, for family members, for spaces, for nonhuman copresences, for issues, for principles, for places, for all kinds of stuff – should challenge many adults (ourselves included) to figure out some new ways of imagining, writing about, relating to and working with children and young people (see also Weller 2007, Harris et al. 2010, Day and Evans 2015). All those normative popular, media and political discourses which figure children and young people solely as dependent *care-recipients*, or as apathetic and *uncaring*, or as folk 'devils' *to-be-feared* 'angels' *to-be-feared-for* … They simply don't give the full picture. Children and young people *do care*, and it is simply not accurate or adequate or fair to think otherwise. We also share Ruth Evans' (2012) sense that noticing everyday, embodied, relational care practices marks a challenge to developmental categories and logics of childhood and youth. Evans (2012) notes how, in case studies of sibling care in households affected by HIV/AIDS in Tanzania and Uganda, the regular, ongoing, repetitive intimate forms of caring done, on-and-on-and-on, by children and young people every day do not neatly fit with linear developmental presumptions of lifecourse transition. We suggest that many chapters in this book can be read as posing a similar kind of challenge by making plain children and young people's wide-ranging participation and reciprocity in caring relationships and communities; their presence and agency in many politicised spaces of care; and their (or rather *our*) co-presence and often careful input with/in many spaces of research and practice. And, in so doing, the chapters collected in this book challenge us (as adults working with children and young people in diverse spaces and roles) to move from a habit of caring for/about children and young people to an ethic of *caring with* children and young people in diverse, plural *caringscapes*.

Structure and contributions of book

As a contribution to these debates the following collection of chapters is structured around three key, overlapping themes. The first group of chapters presents a range of distinctive contributions to the conceptualisation of childhood, youth and care, extending the discussion in Chapter 1 via sustained, substantive engagements with feminist ethics of care and Bourdieusian accounts of habitus, anthropological and psychological theories of kinship, postcolonial theory, critical pedagogies and theories of citizenship, and Foucauldian theories of power, respectively. We hope that this juxtaposition of diversely-theorised chapters will animate broader consideration of the theorisation of care itself, in relation to childhood and youth. In Chapter 2, Gill Hughes thinks through feminist ethics of care in relation to Bourdieu's theories of habitus and 'doxa', and discusses empirical research with young people and practitioners involved in a programme of activities for young people Not in Education, Employment and Training provision ('NEET') in northern England. She argues that a key, often-overlooked feature of care is (in Bourdieu's terms) its 'doxicity': that is, how it often seems 'natural', 'self-evident', or 'understood without question'. Hughes draws upon qualitative research with young people to diagram the nexus of multiple signifiers, dispositions and spatial practices which are understood as care in this particular geographical-historical-political context. Importantly, too, she considers the latent 'darker side' of care: for example, its centrality to many normative, hegemonic discourses, and its constitutive role in forms of guilt, coercion, carelessness or 'overbearing' social relations. Through these reflections, Hughes argues that ethics of care must pay closer attention to context and spatialities. In particular, she calls for greater attention to the ways in which the everyday caring geographies of children, young people and practitioners are framed, constrained and disconnected by contexts of neoliberalisation, marketisation and welfare reforms. In Chapter 3, Rebecca Pratchett and Paul Rees work with a parallel, and in many ways quite separate, lineage of theory on childhood, youth and care, engaging with biological, sociological, ethological and psychological concepts of kinship. They usefully historicise how these intertwining methodological and conceptual traditions have constituted an influential and efficacious set of understandings of care. For example, they propound Bronfenbrenner's Ecological Systems theory as one important way in which the contextuality and spatiality of care has been diagrammed in this context. Some readers (perhaps especially those who have – like us – been schooled in feminist, poststructuralist or social constructionist social sciences) may be discomfited by the developmental and apparently universalising logics that underpin such theories of care. However, we urge readers to approach all chapters with a spirit of openness to interdisciplinary dialogue. Our juxtaposition of diversely-theorised work throughout this book is intended to challenge readers to think about how such different ways of knowing, writing and researching care might be brought into productive interdisciplinary conversations in future work.

In Chapter 4, Sarah de Leeuw considers and historicises the concepts of care which underpin Canadian child welfare policies, particularly as they pertain to contemporary Indigenous families in British Columbia. She traces historical, recent and often-tragic evidence regarding the removal of children from Indigenous families into forms of state 'care'. de Leeuw argues that this kind of state intervention – in the interests of 'child protection' and 'care' – can only be understood with reference to postcolonial critiques of hegemonic colonial systems and logics. In the British Columbian case, she evidences how legal, political and popular discourses of Indigenous communities as 'others' in need of special protection and transformation have become so naturalised as to appear 'common sense'. As such, Chapter 4 highlights the need for theorisations of childhood, youth and care to be carefully attuned to historical-political contexts, and always already critical of 'common sense' understandings of care. As de Leeuw shows, the consequences of 'common sense' can too often be tragic, and well-intentioned forms of 'care' can often produce or legitimise significant, enduring forms of harm. In Chapter 5, Bronwyn Wood and Rowena Taylor consider another widely significant (and, again, seemingly 'common sense') articulation of 'care' in relation to childhood and youth: namely, the global preponderance of policies and practices which promote or require 'active citizenship' (and, therefore, more active forms of care) on the part of children and young people. Drawing on research in New Zealand schools, Wood and Taylor explore how teachers and young people diversely relate to initiatives that effectively require them to care *more* about particular issues, communities and spaces. In so doing they illustrate the multiple ways in which emotions of care (or *not-caring*) are constituted within formal spaces of education, and in policy initiatives that encourage young people to engage in public service, civic participation, social action and volunteering. They also challenge readers to be more critical and creative in considering interconnections between emotions of caring, acts of citizenship, and formal educational and citizenship initiatives.

In Chapter 6, Tom Disney offers another productive way of theorising childhood, youth and care via a substantial engagement with Michel Foucault's writing on biopolitics. He argues that concepts of biopolitics allow for a fuller appreciation of the material, embodied, and bio-social constituents of childhood, care and care-itself. Disney's chapter both introduces Foucault's work on biopolitics and suggests that extant social scientific engagement with this work have tended to be somewhat limited and uncritical, particularly in relation to their spatiality. To illustrate the complex interrelations between childhood, care, biopolitics and spatiality,

Disney reviews a range of qualitative research on orphanage spaces. Through a heart-rending discussion of these institutional spaces, he highlights the complex, multiscalar way in which biopower operates within orphanage spaces. In so doing, he calls for further research on spatial forms of power and control within diverse institutional spaces that are designed to offer 'care' for children and young people. Disney's chapter also forms part of a second group of chapters which present new research focusing upon children and young people in, or in relation to, formal designated spaces or institutions of care. Chapters 6–8 specifically consider orphanages, secure care units and early childhood centres. However, we hope that these empirical discussions will also prompt further reflection and research to explore children and young people's presence in, and relationships to a wider range of diverse spaces of care. As these chapters show, such spaces can impact profoundly – positively or negatively – upon the lives of children and young people. For example, in Chapter 7 Anna Schliehe and Annie Crowley consider the experiences of young women in Scottish 'secure care' units, for individuals deemed to pose a significant risk to themselves or others. Through their deeply-affecting research Schliehe and Crowley highlight some of the practical, legal and spatial tensions which characterise institutions that offer a 'place of safety' via tightly controlled institutional logics and procedures. Drawing on a rich range of research methods, they juxtapose the perspectives of young women and staff in 'secure care' settings and highlight young people's complex pathways through intensely-regulated penal and welfare institutions. They thus call for more careful research to consider the intersections of children and young people's lifecourses with changing policy-legislative spaces of penal-welfare care in diverse contexts. Drawing upon research with young children in New Zealand, Ann Pairman and Carmen Dalli consider the ways in which care is constituted in the quite different (but also intensely spatially regulated) contexts of early childhood centres. In considering children's experiences of these designated spaces of play, care, family support and early childhood learning, Pairman and Dalli highlight the ways in which children themselves are not just recipients of parental, service-providers' or state care; instead they are active participants in the co-construction of care with peers, and with the affordances of complex built environments. Pairman and Dalli thus draw attention to the complex ways in which children and young people engage with the spaces, environments and materialities of spaces of care – and they call for new ways of theorising children's agency, mobility, and social interactions which acknowledge such complexity and co-construction.

A third group of chapters feature empirical case studies of children and young people as carers or recipients of care, in diverse contexts. These chapters prompt reflection upon the many kinds of caring practices and roles which (contra many contemporary discourses of children and young people as uncaring, apathetic and/or as passive recipients of care) are undertaken by many children and young people. For example, in Chapter 9 Caroline Day presents new research on young people's caring roles in Zambia, in the context of the HIV/AIDS epidemic in southern Africa. Day's deeply affecting research shows that, although young people are increasingly called upon to provide a significant array of everyday, medical, emotional and economic support for chronically-ill family members in this context, this work is rarely labelled or acknowledged as care. Indeed, it is notable that the young people consulted in Day's study do not, themselves, identify with the role of 'caregiver', but see their significant responsibilities as just 'what is expected'. Day develops an argument about the difficulty of characterising and narrating young people in terms of their caring responsibilities, when their caregiving is so seldom represented or valued within many contemporary social-cultural contexts. As previously noted, Ruth Evans' research on family caring practices in East and West Africa has been important in acknowledging and theorising children and young people's care. In Chapter 10 she extends this work by presenting research about young people's responses to the death of a parent or carer. In relation to studies of HIV/AIDS epidemics in the global south, Evans notes that, despite a significant literature on mortality and orphanhood, research has tended to diagnose material and social consequences of orphanhood rather than engaging with children and young people's everyday, emotional and familial experiences parental death. Through a moving discussion of qualitative research with orphaned young people in Tanzania, Uganda and Senegal, she highlights many ways in which children and young people perform caring practices which sustain connections with family members and enact 'continuing bonds' with deceased family members. Through this empirical material, Evans argues that research on children, young people and care should pay closer attention to experiences of bereavement. She also highlights the significance of social-relationalities (and, specifically 'sibling practices') in constituting everyday lives and caring practices in diverse contexts.

In Chapter 11, Tracy Hayes invites readers to reflect upon incidents from research with 'looked after' young people participating in organised activities designed to engage them in outdoor and 'natural' spaces in the UK. Fundamentally, Hayes challenges us to consider the diverse ways in which people may experience and *care about* a particular space, issue or experience. For example, she notes that young people may care about outdoor walking experiences in ways which differ markedly from the expectations of adult organisers of projects for young people. Her chapter therefore considers the ways in which children and young people care about – and enact multiple forms of *kindness* towards – themselves, others, co-presences, landscapes and natures. Hayes also opens out a series of challenging questions about the extent to which some predominant habits of academic research, writing and thinking may not always be conducive to *caring scholarship*. Her chapter is an invitation and a plea for researchers to slow down and to pause to reflect how we might work in more *caring, careful*, moral and equitable ways. In Chapter 12, Sam Wilkinson details the *caring, careful* (and occasionally *careless*) practices of young women consuming alcohol in north-west England. She argues that research on children, young people and care has tended to be relatively limited in its spatial and relational foci (tending to focus upon familial care within the home), and in its implicit reproduction of normative representations of care (often figuring young people as uncaring, or as prone to careless, risky behaviour).

Against this grain, Wilkinson's chapter highlights the every-night caring practices of friendship groups of young women engaged in drinking practices in a range of spaces, including diverse outdoor drinkscapes. She notes how caring practices in such contexts may be pre-planned, contingent or spontaneous; the chapter is also important in highlighting some ways in which digital communications technologies may supplement 'face-to-face' forms of care. In so doing, Wilkinson develops a wider call for further research exploring emotional, affective and embodied characteristics of children and young people's caring practices, whilst critiquing normative representations of care.

Chapter 13 – by Sally Robinson, Karen Fisher, Malcolm Hill and Anne Graham – is significant in highlighting the caring relationships and practices of children and young people with diverse disabilities. They note that people with disabling conditions are too often considered through a normative lens that positions them as inevitably *care-recipients, in-need-of-care*. Through their chapter they highlight more critical, affirmative theorisations of disability and care that recognise caring relationalities and mutualities. However, they note that relatively little research has considered how children and young people with disabilities experience, and participate in, caring relationships and interventions. Robinson et al. draw upon participatory research with a sample of Australian young people with diverse cognitive disabilities. They evidence how many of these participants were active agents in multiple forms of reciprocal caring relationships, and note how different kinds of support services could facilitate or circumscribe these relationships. They particularly call for policy-makers and service providers to constitute further opportunities for disabled children and young people to develop their agency in these contexts. In Chapter 14, Karen Wells explores the lifecourses, mobilities and caring relationships of children who are privately fostered across international borders. Through research with young migrants from Bangladesh to the UK, she explores the complexity and multidimensionality of care, as children move to live with distant relatives or 'strangers'. As she explores the forms and motivations of this mode of transnational care, Wells is attentive to the experiences and identities of children and young people who are privately fostered: in particular she notes the complex interplays between religions, cultural 'obligations', national identities and belonging within the 'stretched' and extensive networks of private fostering arrangements.

We are sure that this collection of chapters will stimulate wide-ranging reflection, discussion and questioning around the theme of childhood, youth and 'care'. To close the book, we present three reflective pieces by discussants who have engaged with this edited collection. First, Kathrin Hörschelmann (Chapter 15) lays out a series of challenging conceptual ambivalences prompted by the book's chapters. For example, she challenges readers to critique (our) normative, naturalised assumptions about care and responsibility but not allow scepticism about care to become *the-reverse-of-care*; to problematise normative discourses of children and young people as vulnerable recipients of care but not abrogate adult responsibilities to children and young people; to develop more caring practices but also think about our inability to 'care for everyone/everything'; and to 'be prepared to listen to accounts by/of parents and other adults who do not care, who care differently, or who abuse children'. Second, Michelle Pyer (Chapter 16) considers the implications of the book's chapters for policy-makers, practitioners, institutions and service providers who work with children and young people. Second, we (Pyer and Horton, Chapter 16) consider the implications of the book's chapters for policy-makers, practitioners, institutions and service providers who work with children and young people. We highlight a series of evidence-based messages which recur throughout the book and use these to develop a set of recommendations for future practice for spaces of care, and more generally for those seeking to work in more caring ways with children and young people. Finally, Sophie Bowlby (Chapter 17) offers a concluding reflection upon spatialities of childhood, youth and care, as manifest so vividly in this book's chapters. As previously mentioned, Bowlby's work on caringscapes and carescapes has been significant in developing conceptualisations of care (and indeed these concepts underpin many chapters in this book) so we are really grateful for her reflective piece upon children and young people's presence and participation within caringscapes.

References

Allen, C. (2001) Back to the old house? 'Sick role' and biographical narratives of the housing needs of short stay hospital patients. *Health and Place* 7, pp.81–92.

Amin, A. (2002) Ethnicity and the multicultural city: living with diversity. *Environment and Planning A* 34, pp.959–80.

Armstrong, P. and Armstrong, H. (2001) *Thinking it through: women, work and caring in the new millennium*. Halifax, Healthy Balance Research Program.

Atkinson, S., Lawson, V., and Wiles, J. (2011) Editorial. Care of the body: spaces of practice. *Social and Cultural Geography* 12, pp.563–72.

Barnes, M. (1997) *Care, communities and citizens*. London, Longman.

Barnes, M. (2012) *Care in everyday life: an ethic of practice*. Bristol, Policy Press.

Barnett, C. (2005) Ways of relating: hospitality and the acknowledgement of otherness. *Progress in Human Geography* 29, pp.2–21.

Bartos, A. E. (2012). Children caring for their worlds: the politics of care and childhood. *Political Geography* 31, pp.157–66.

BBC (2016) Young people don't care about politics … or do they? Available at www.bbc.co.uk/news/uk-politics-eu-referendum-35775773

Bell, D. (2007) The hospitable city: social relations in commercial spaces. *Progress in Human Geography* 31, pp.7–22.

Boserup, E. (1970) *Women's role in economic development*. London, Allen and Unwin.

Bowlby, S. (2011) Friendship, co-presence and care: neglected spaces. *Social and Cultural Geography* 12, pp.605–22.

Bowlby, S. (2012) Recognising the time-space dimensions of care: caringscapes and carescapes. *Environment and Planning* A 44, pp.2101–18.

Bowlby, S., McKie, L., Gregory, S. and MacPherson, I. (2010) *Interdependence and care over the lifecourse*. London, Routledge.

Brechin, A., Walmsley, J., Katz, J. and Peace, S. (1999) *Care matters*. London, Sage.

Brickell, K. and Datta, A. (eds.) (2011) *Translocal geographies: spaces, places, connections*. Farnham, Ashgate.

Brinegar, S. (2003) The social construction of homeless shelters in the Phoenix area. *Urban Geography* 24, pp.61–74.

Brown, M. (2003) Hospice and the spatial paradoxes of terminal care. Environment and Planning A 35(5), pp.833–851.

Cloke, P., May, J. and Johnsen, S. (2008) Performativity and affect in the homeless city. *Environment and Planning D: Society and Space* 26, pp.241–63.

Cloke, P. and Beaumont, J. (2013) Geographies of postsecular rapprochement in the city. *Progress in Human Geography* 37, pp.27–51.

Conradson, D. (2003a) Spaces of care in the city: the place of a community drop-in centre. *Social and Cultural Geography* 4, pp.507–24.

Conradson, D. (2003b) Editorial. Geographies of care: spaces, practices, experiences. *Social and Cultural Geography* 4, pp.41–454.

Corcoran, A. (2011) Caregiving styles: a cognitive and behavioural typology associated with dementia family caregiving. *The Gerontologist* 51, pp.463–72.

Datta, K., McIlwaine, C., Evans, Y., Herbert, J., May, J. and Wills, J. (2010) A migrant ethic of care? Negotiating care and caring among migrant workers in London's low-pay economy. *Feminist Review* 94, pp.93–116.

David, T., Goouch, K. and Powell, S. (eds.) (2016) *The Routledge international handbook of philosophies and theories of early childhood education and care*. London, Routledge.

Day, C. (2016) Education and employment transitions: the experiences of young people with caring responsibilities in Zambia. In Abebe, T., Waters, J. and Skelton, T. (eds.) *Geographies of children and young people volume 10: labouring and learning*. New York, Springer.

Day, C. and Evans, R. (2015) Caring responsibilities, change and transitions in young people's family lives in Zambia. *Journal of Comparative Family Studies* 46, pp.137–52.

Daya, S. and Wilkins, N. (2013) The body, the shelter, and the shebeen: an affective geography of homelessness in South Africa. *Cultural Geographies* 20, pp.357–78.

Dermott, E. and Seymour, J. (eds.) (2011) *Displaying families*. London, Palgrave.

Dyck, I., Kontos, P., Angus, J. and McKeever, P. (2005) The home as a site for long-term care: meanings and management of bodies and spaces. *Health and Place* 11, pp.173–85.

Dyer, S., McDowell, L. and Batnitzky, A. (2008) Emotional labour/body work: the caring labours of migrants in the UK's National Health Service. *Geoforum* 39, pp.2030–8.

England, K. (2010) Home, work and the shifting geographies of care. *Ethics Place and Environment* 13, pp.131–50.

Engster, D. and Hamington, M. (eds.) (2015) *Care ethics and political theory*. Oxford, Oxford University Press.

Evans, B. (2010) Anticipating fatness: childhood, affect and the pre-emptive 'war on obesity'. *Transactions of the Institute of British Geographers*, 35, pp.21–38.

Evans, B., Colls R., and Hörschelmann, K. (2011) 'Change4Life for your kids': embodied collectives and public health pedagogy. *Sport, Education and Society* 16, pp.323–41

Evans, R. (2010) Children's caring roles and responsibilities within the family in Africa. *Geography Compass* 4, pp.1477–96.

Evans, R. (2012) Sibling caringscapes: time-space practices of caring within youth-headed households in Tanzania and Uganda. *Geoforum* 43, p.824–35.

Evans, R. (2015) Negotiating intergenerational relations and care in diverse African contexts. In Vanderbeck, R. and Worth, N. (eds.) *Intergenerational Space*. London, Routledge, pp.199–213.

Evans, R. and Becker, S. (2009) *Children caring for parents with HIV and AIDS: global issues and policy responses*. Bristol, Policy Press.

Fisher, B. and Tronto, J. (1990) Towards a feminist theory of caring. In Abel, E. and Nelson, M. (eds.) *Circles of care: work and identity in women's lives*. Albany, State University of New York Press, pp.35–62.

Gardiner, J. (1997) *Gender, care and economics*. London, MacMillan.
Glazer, N. (1984) Servants to capital: unpaid domestic labour and paid work. *Review of Radical Political Economics* 16, pp.60–87.
Gleason, T. and Narvaez, D. (2014) Child environments and flourishing. In Narvaez, D., Valentino, K., Fuentes, A., McKenna, J and Gray, P. (eds.) *Culture, childrearing and social wellbeing*. Oxford, Oxford University Press, pp.335–48.
Green, M. and Lawson, V. (2011) Recentring care: interrogating the commodification of care. *Social and Cultural Geography* 12, pp.639–54.
Hall, E. (2011) Shopping for support: personalisation and the new spaces and relations of commodified care for people with learning disabilities. *Social and Cultural Geography* 12, pp.589–603.
Harris, A., and Wyn, J. (2009) Young people's politics and the micro-territories of the local. *Australian Journal of Political Science* 44, pp.327–44.
Harris, A., Wyn, J., and Younes, S. (2010) Beyond apathetic or activist youth: 'ordinary' young people and contemporary forms of participation. *Young* 18, pp.9–32.
Held, V. (2006) *The ethics of care: personal, political, and global*. Oxford, Oxford University Press.
HSCIC [Health and Social Care Information Centre] (2014) NHS workforce: summary of staff in the NHS. London, HSCIC.
Huang, S., Yeoh, B. and Toyota, M. (2012) Caring for the elderly: the embodied labour of migrant care workers in Singapore. *Global Networks* 12, pp.129–34.
James, A. and James, A. (2004) *Constructing childhood: theory, law and social practice*. Palgrave, London
Jones, R., Pykett, J. and Whitehead, M. (2013) *Changing behaviours: on the rise of the psychological state*. Cheltenham, Edward Elgar.
Kearns, R. and Joseph, A. (2000) Contracting opportunities: interpreting post-asylum geographies of mental healthcare in Auckland, New Zealand. *Health and Place* 6, pp. 159–70.
Kraftl, P., Horton, J. and Tucker, F. (2014) Children's geographies: an annotated bibliography. *Oxford Bibliographies*. Available at www.oxfordbibliographies.com/view/document/obo-9780199791231/obo-9780199791231-0080.xml
Lambie-Mumford, H. and Green, M. (2015) Austerity, welfare reform and the rising use of food banks by children in England and Wales. *Area* online early. http://onlinelibrary.wiley.com/doi/10.1111/area.12233/full (accessed 21 February 2017).
Laurier, E. and Philo, C. (2006) Cold shoulders and napkins handed: gestures of responsibility. *Transactions of the Institute of British Geographers* 31, pp.193–207.
Laurier, E., Whyte, A. and Buchner, K. (2002) Neighbouring as an occasioned activity: 'finding a lost cat'. *Space and Culture* 5, pp.346–67.
Lawson, V. (2007) Geographies of care and responsibility. *Annals of the Association of American Geographers* 97, pp.1–11.
McDowell, L. (2004) Work, workfare, work/life balance and an ethic of care. *Progress in Human Geography* 28, pp.145–63.
McEwan, C. and Goodman, M. (2010) Place geography and the ethics of care: introductory remarks on the geographies of ethics, responsibility and care. *Ethics, Policy and Environment* 13, pp.103–12.
McKie, L. and Bowlby, S. (2004) *Caringscapes: experiences of caring and working*. Edinburgh, Centre for Research on Families and Relationships.
Milligan, C. (2000) Bearing the burden: toward a restructured geography if caring. *Area* 32, pp.49–58.
Milligan, C. and Wiles, J. (2010) Landscapes of care. *Progress in Human Geography* 34, pp.736–54.
Nicholson, L. (1988) Gender and history: the limits of social theory in the age of the family. New York, Columbia University Press.

Noddings, N. (2013) *Caring: a relational approach to ethics and moral education.* Berkeley, University of California Press.
ONS [Office for National Statistics] (2013) *Women in the labour market*. London, ONS.
Parr, H. (2000) Interpreting the hidden social geographies of mental health: ethnographies of inclusion and exclusion in semi-institutional places. *Health and Place* 6, pp.225–37.
Parr, H. (2003) Medical geography: care and caring. *Progress in Human Geography* 27, pp.225–37.
Parr, H. (2008) *Mental health and social space: towards inclusionary geographies*. Oxford, Blackwell.
Philo, C. (1997) Across the water: reviewing geographical studies of asylums and other mental health facilities. *Health and Place* 3, pp.73–89.
Philo, C. (2004) *A geographical history of institutional provision for the insane from Medieval times to the 1860s in England and Wales: the space reserved for insanity*. Lewiston, Edwin Mellen.
Pykett, J. (ed.) (2011) *Governing through pedagogy: re-educating citizens*. Routledge, London.
Pykett, J. (2015) *Brain culture: shaping policy through neuroscience*. Bristol, Policy Press.
Ranghuram, P., Madge, C. and Noxolo, P. (2009) Rethinking responsibility and care for postcolonial worlds. *Geoforum* 40, pp.5–13.
Robson, E. (2004) Hidden child workers: young carers in Zimbabwe. *Antipode* 36, pp. 227–48.
Rogers, C. and Weller, S. (eds.) (2013) *Critical approaches to care: understanding caring relations, identities and cultures*. London, Routledge.
Rossi, A. (ed.) (2001) *Caring and doing for others: social responsibility in the domains of family, work and community*. Chicago, University of Chicago Press.
Rowland, E. (2014) *Emotional geographies of care work in the NHS*. Unpublished doctoral thesis. London, King's College.
Sevenhuijsen, S. (1998) *Citizenship and the ethics of care*. London, Routledge.
Silk, J. (1998) Caring at a distance. *Philosophy and Geography* 1, pp.165–82.
Silk, J. (2004) Caring at a distance: gift theory, aid chains and social movements. *Social and Cultural Geography* 5, pp.229–51.
Skelton, T. (ed.) (2016) *Geographies of children and young people: major reference work*. New York, Springer.
Skills for Care (2015) *The state of the adult social care sector and workforce in England*. London, Skills for Care.
Smith, S. (2005) States, markets and an ethic of care. *Political Geography* 24, pp.1–20.
Staeheli, L. and Brown, M. (2003) Where has the welfare gone? Introductory remarks to the geographies of care and welfare. *Environment and Planning A* 35, pp.771–7.
Thomas, C. (1993) De-constructing concepts of care. *Sociology* 27, pp.649–69.
Tronto, J. (1993) *Moral boundaries: a political argument for an ethic of care*. London, Routledge.
Tronto, J. (1989) Women and caring: what can feminists learn about morality from caring. In Jaggar, M. and Bordo, S. (eds.) *Gender/body/knowledge: feminist reconstructions of being and knowing*. New Brunswick, Rutgers University Press, pp.172–87.
Tronto, J. (2013) *Caring democracy: markets, equality and justice*. New York, NYU Press.
Valentine, G. (1996) Angels and devils: the moral landscape of childhood. *Environment and Planning D* 14, pp.581–99.
Valentine, G. (2008) Living with difference: reflections on geographies of encounter. *Progress in Human Geography* 32, pp.323–37.
Vivat, B. (2002) Situated ethics and feminist ethics in a west of Scotland hospice. In Bondi, L., Avis, H., Bankey, R., Bingley, A., Davidson, J., Duffy, R., Einagel, V., Green, A., Johnston, L., Lilley, S., Listerborn, C., Marshy, M., McEwan, S., O'Connor, N., Rose, G., Vivat, B. and Wood, N. (eds.) *Subjectivities, knowledges, and feminist geographies*. Oxford, Rowman and Littlefield, pp.236–52.

Weller, S. (2007) *Teenagers' citizenship: experiences and education*. London, Routledge.
White House (2015) *Press release: helping all working families with young children afford child care*. Available at www.whitehouse.gov/the-press-office/2015/01/21/fact-sheet-helping-all-working-families-young-children-afford-child-care
Wiles, J. (2011) Reflections on being a recipient of care: vexing the concept of vulnerability. *Social and Cultural Geography* 12, pp.573–88.
Williams, A. (2002) Changing geographies of care: employing the concept of therapeutic landscapes as a framework in examining home space. *Social Science and Medicine* 55, pp.141–54.
Williams, A. (ed.) (2007) *Therapeutic landscapes*. Farnham, Ashgate.
Williams, J. (1996) Geography of the homeless shelter: staff surveillance and resident resistance. *Urban Anthropology* 25, pp.75–113
Wood, B. (2013) Young people's emotional geographies of citizenship participation: spatial and relational insights. *Emotion, Space and Society* 9, pp.50–8.
Wood, B. (2016) Excluded citizens? Participatory research with young people from a 'failing' school community. *Children's Geographies* 14, pp.310–24.
Young, I. M. (1981) Beyond the unhappy marriage. A critique of the dual systems theory. In Sargent, L. (ed.) *Women and revolution*. Boston, South End, pp.43–69.
Zhang, Y. and Jean Yeung, W. (2012) Shifting boundaries of care in Asia: an introduction. *International Journal of Sociology and Social Policy* 32, pp.612–22.

2 Who cares and how would you know? Conceptualising ‘doxic’ care

Gill Hughes

Introduction

> “… it is hard to explain without using the word care … well … they do care … I can’t explain it …”

This was the response from a participant, Helena aged 18, in the research study discussed in this chapter; as will be seen she was not alone in this quandary. Care is something which seems natural, almost invisible, yet its self-evidence problematizes its definition. This chapter argues that care contains ‘doxic’ qualities (Bourdieu 1977). That is, its meaning is taken as understood without question, which problematises how someone might define care. Thus it is necessary to explore various conceptualisations of care, whilst recognising that these are but a few within vast and increasing discourse, but essential in order to understand how care manifests as a catalyst to facilitate school connectedness.

The complexity of defining care renders it a subjective and contested concept impacting both the carer (the one who ‘does’ the caring) and the ‘caree’ (the one who ‘receives’ care). Misrecognition can take place, particularly when a mismatch occurs between what is intended as care and how it is received; for example, when one person’s caring act might be perceived by another as surveillance. To assist in the process of understanding care, illustrations will be drawn from a doctoral study set in Northwick, a northern town in England, exploring young people, care and (dis)connectedness within spaces of education. The chapter draws on interviews with young people attending Not in Education, Employment and Training provision (NEET), adult students, teachers, alternative educators and those providing ‘turnaround’ experiences, designed to re-connect young people to education.

This chapter argues for centralising context in order to understand the lived realities of young people, and the practitioner/carer’s milieu in the field of education. This assists when identifying what might enhance or impede the ability to care and thus connect. One particular constraint, explored below, is framed within a landscape of neoliberalism, manifest as market-driven welfare services introduced from the late 1970s in the UK, infused with a culture of targets and, more recently, austerity cuts. The privileging of profit over people increasingly renders care as a commodity to be calculated, packaged and provided, far from its perceived natural supportive characteristics (Moss 2014a).

The chapter begins with an overview of connectedness, then develops a synopsis of care literature, before concluding with a call for new vistas to centralise care in theory and practice to underpin politics, policy, practice and relationships, arguing for safe spaces for practitioners to examine the doxic construct of care.

Setting the context: discovering care in connectedness

Connectedness, as a construct, is less familiar in the UK, and mainly utilised in America (US) and Australia. The *Wingspread Declaration on School Connections* was developed in 2003 at a conference sponsored by the US Centers for Disease Control and Prevention (CDC), Division of Adolescent and School Health. The *Declaration* (Wingspread Conference 2004, p.231) defined connectedness as the "belief by students that adults in the school *care about their learning and about them as individuals*" (added emphasis); much of the associated literature originates within the fields of 'adolescent and school health' (Barber and Schluterman 2008, p.209). Chapman et al (2013) contend that,

> the school social context is critical in shaping adolescent behaviour. School connectedness, in particular, has been repeatedly identified as an important protective factor. It has been shown to be positively associated with school retention and emotional health and wellbeing, and negatively associated with adolescents' involvement in risk taking behaviours. (p.95)

This is crucially facilitated by a significant adult, an educational practitioner, who performs care to secure connection to education and school, generating positive outcomes for young people. In England the closest construct would be inclusion, which dominates educational discourse informing policy on engagement *for all* in mainstream schools. Dunne (2008, p.1), taking a Foucauldian view, posits that inclusion is taken as 'self-evident', similar to the argument here on care; without deconstruction inclusion can manifest as a 'universalizing' concept engendering "a potentially normalizing, hegemonic discourse". Dunne identifies that inclusion has often been generally synonymous with special educational needs (SEN). This narrow view is problematic, although authors such as Booth and Ainscow (1996) have continuously sought to expand the definition wider than SEN, taking account of lived experiences, such as poverty and deprivation. Connectedness is therefore not without issues, although the literature has often been little critiqued because its aims, like those of inclusion, seem rational. Moreover, the underlying intent of interventions can be read beyond the outcomes above, rather to ensure young people become fiscally active future citizens, as illustrated by Fernandez and Gabe (2009) in the US, which similarly befits the English neoliberal ambition to prevent school leavers transitioning as welfare claimants (Chakelian 2015). In this sense connectedness could fit Dunne's earlier caution on generating normativity.

In addition, the relational aspect suggests that connection can change rapidly to disconnection where misunderstandings of what constitutes care impacts relationships, hence the placement of (dis) before connectedness signifying precarity through the ease of rupture; it is thereby all the more urgent to understand the manifestations of care in its role as catalyst for connectedness (see also Hughes 2015). However, whilst mindful of the issues, the emergent notion of connectedness could provide an opportunity to explore inc/exclusion within school through a different lens offering potential as a heuristic device. The term suggests a two-way process where either party can connect or disconnect, arguably offering a degree of power to young people not normally afforded through deficit labels such as excluded, disaffected, disadvantaged or marginalised. Disconnection reframed could posit a site of resistance to normativity and conformity where the universality of education systems do not meet needs therefore, providing an active choice not to connect, through non-attendance or not engaging in class (see also Hughes 2015). This process would currently be problematised, committing to a re-imagined use of connectedness positions young people as 'freer than they feel' (Foucault 1988, p.10), utilising agency to resist the status quo. This should prompt an exploration of the system and its dysfunction as opposed to labelling young people as problematic. Nonetheless, connectedness requires care to fulfil its potential, therefore an examination of the construct of care is crucial if this development is to be pursued.

Care literature in synopsis

As outlined in Chapter 1, relevant care literature spans a wide variety of disciplines. Care can be examined in theory (Fisher and Tronto 1990, Tronto 1993, 2013, 2015, Sevenhuijsen 1998, Held 2006, Hollway 2006) or in practice (van Manen 1990, 2002, Barnes 1997, 2012, Noddings 2002, 2003, 2013, Brannen and Moss 2003, Petrie 2003, Fine 2007, Rogers and Weller 2013, Moss 2014a, 2014b). Care is also explored within particular paradigms such as the 'ethic of care' (EoC) which influenced the development of the discourse of care from the 1980s, originating with Gilligan (1982) and Noddings (1984) (see Chapter 1). The EoC foregrounded the necessity to replace traditional, rational and universalising approaches to justice, apparent within Kantian (1724–1804) duty-based deontological ethics and Bentham's (1748–1832) Utilitarian principles, with the ethics of care focusing on relationships (Banks et al 2013). The difference is explicated as,

> an ethic of justice focuses on questions of fairness, equality, individual rights, abstract principles, and the consistent application of them. An ethic of care focuses on attentiveness, trust, responsiveness to need, narrative nuance, and cultivating caring relations, whereas an ethic of justice seeks a fair solution between competing individual interests and rights, an ethic of care sees the interests of carers and cared-for as importantly intertwined.... Whereas justice protects equality and freedom, care fosters social bonds and co-operation. (Held 2006, p.15)

Early EoC theorists offered important recognition and validation to the contribution that women made in the field of care, with the mother–child relationship central to understanding the manifestation of care in practice, whilst claiming this can be applicable to men too (see Barnes 2012 for an overview). The caution is to avoid essentialism by generating a synonymous link between care and women/mothers, reinforcing traditional gender roles (Moss 2014a; Barnes 2012; Butler cited in Hark and Villa 2011). This risks marginalising its importance as 'women's issues', suggesting only partial concern (Barker 2014), but also identifying the problematics of masculinity juxtaposed with care for male teachers (Hansen and Mulholland 2005). In education it also impacts the perception and experience of 'non-parent' teachers, as in childless women (Vogt 2002: 261). However, Philip et al (2013) note that:

> feminist conceptions of care can challenge persisting assumptions about human relationships, decision-making and allocation of resources; including time, emotions, love and labour, in personal, social and institutional contexts. This is important in order to avoid limiting care as either a taken-for-granted idea or as a professionalised term to describe a commercial service. (p.3)

As the EoC developed, a merger of care and justice unfolded (Tronto 2013, 1993, Sevenhuijsen 1998). Held (2006, p.15) cites Sara Ruddick's (1996) contention that "justice is always seen in tandem with care". The importance of this theoretical discourse is the focus on the necessity to critically reflect upon care, to understand it relationally and within context. Brannen and Moss (2003) suggest that the

> ethics of care provide a different way of thinking about care … [it] moves us from care as a task performed by adults on children to care within relational encounters of all forms, which suggests a two-way process, linking it to the potential of connectedness. (p.39)

The depth of care

Philip et al (2013) suggest that "care can do important conceptual work", drawing out some key points, which equally underscore the discussions here:

> [there is the] potential of 'care' for responding to pertinent current issues such as social justice, exclusion and connection [and suggest it necessary to] think [...] about and with 'care' [in a] [...] more moral and philosophical direction. (pp.1–2)

Care, as a lynchpin for educational connectedness, is predicated on the ethos and ontological practices of the school, which are influenced by wider contextual components such as ideological framings within government policy. Noddings' (2002) conceptualisation of care is relational. It requires a circularity, where the carer acknowledges that the caree is in need of care, referred to as 'receptive attention'. The carer offers 'care'; this is '*motivational displacement*' (original emphasis). The carer's 'motive energy begins to flow to' the caree, and if the caree acknowledges the receipt of the gesture as care, they convey this to the carer, resulting in acceptance that care has taken place. If this process falters (if the caree does not acknowledge the gesture as care) no caring encounter occurred. The latter suggests possibilities for misrecognition, mismatching intent and receipt, and problematically generating the 'dis' before connection between carer and caree. Such subjectivity is drawn out here:

> 'Care' is one of those words widely used yet rarely defined. As a result "the concept of care is both ambiguous and contested ... and has been used in such diverse ways that it is in danger of losing its core meaning". (Daly and Lewis 1999, cited in Brannen and Moss 2003, p.4)

Thus care's constant usage can appear with little consideration of meaning. This complexity is illustrated with an alternative education provider's response to a request to define care in the Northwick study:

> Care? [Pause] What's my definition of care? That's ... I think my immediate response is that it is about making a *genuine* connection *with* somebody as an individual and *genuinely* wanting *to* [pause] support them, *I am trying to avoid using the word care* [both laugh].

Rogers and Webb (1991) explicate: "our knowledge of caring is tacit; it is implicit in action. In other words, although we have difficulty defining it, we know it when we see it" (p.177). This definitional difficulty does not only reside with practitioners and the recipients, but also with authors in the field. Held (2006) explored her own issues with defining care "in a precise way":

> [t]here is not yet anything close to agreement among those writing on care on what exactly we should take the meaning of this term to be.... For over two decades, the concept of care as it figures in the ethics of care has been assumed, explored, elaborated, and employed in the development of theory. But definitions have often been imprecise, or trying to arrive at them has simply been postponed (as in my own case).... Some ... have attempted to be precise ... others have proceeded with the tacit understanding ... we know what we are talking about ... [b]ut care has many forms, and as the ethics of care evolves, so should our understanding of what care is. (p.29)

Moss (2014a, p.422) states that "[c]are ... is one of those words ... much used ... less often examined". To counter the difficulty for participants in Northwick when defining care, it was necessary to utilise a phenomenological question – what does it *feel like* to be cared about? van Manen (1990) offers an example relating to learning: phenomenology would ask, not *how* someone learns, but what the experience *feels* like, in context. For the most part therefore, participants in the Northwick study drew out *signifiers of care* (Box 2.1) in lieu of a definition.

Box 2.1 Signifiers of care in Northwick

- Feeling looked-after
- Feeling bothered-about
- Feeling Listened-to
- Feeling safe [a constant referent]
- Feeling valued
- Being interested [in the young person]
- Being known [in context]
- Being funny [also banter/laughter/teasing]
- Being firm-but-fair
- Constructiveness
- Honesty
- Respect
- Consistency
- Interesting and creative teaching
- Being human

It could be suggested that the *signifiers* listed in Box 2.1 point to ways of being when generating connection – with care. Noddings (2003) argues further that

> It is not only students' rationality that must be respected; students need and want teachers to care for them as persons and to convey this care through *listening and responding* to their expressions of concern.... The teacher as person is centrally important in teaching ... the teacher sets an example with her whole self – her intellect, her responsiveness, her humour, her curiosity ... her care. (p.244, added emphasis)

Noddings' implication positions the teacher as lynchpin to the performing of care, requiring the appropriate dispositions in order to do so. Tronto (1993, p.127) contends that care includes "attentiveness, responsibility, competence and responsiveness", which is taken beyond the human to animals and the environment as "a species activity" (see Fisher and Tronto 1990). Brannen and Moss (2003, p.12) similarly equate care with "love, responsibility and attentiveness". Later Tronto (2013, p. x) states "care has many meanings" such as "reaching out to something", or as "actions and disposition"; importantly she suggests that "care expresses relationships", and can demonstrate "deepest convictions ... I care about". Moss (2014b, p.423) is influenced by Tronto's work identifying that "care ... [is] a relational ethic, defining how we should relate to each other and the wider world", which must take account of 'otherness' and diversity. Tronto (1993, p.127) posits that care is "a practice rather than a set of rules or principles". However, the range of *signifiers of care* is problematised by its manifestations as a virtue, feeling, service, activity, skill, duty and commodity (Hughes 2015). Fine (2007) elucidates care's multifunctionality further:

> Care can be understood as a form of *meaningful work*, as *social solidarity*, as an expression of *interpersonal relationships* and as a source of *personal meaning*.

> But it can also be *reduced* again to a *set of household duties* for women, recast as a form of *custodial protection [being in care]*, or fashioned as a *product to be packaged and sold for corporate profit*, its costs kept low by reliance on unskilled and casualized labour and by competition between providers. *A single definition of care remains elusive*, I argue, in part because it refers at once to *an ideal set of* ***values*** *and a series of concrete* ***practices***. (p.2, added emphasis)

These examples demonstrate the complexity in attempting to secure a bounded meaning, leaving care open to subjective interpretation, which could be argued as importantly offering scope for negotiation by preventing a universal approach when negotiating needs, but might equally be ad hoc and open to partiality or coercive practices, thus again illustrating the necessity for clarity within practice. Such subjectivity adds to the complexity, leading to 'doxa', put simply as "the knowledge we think with but not about" (Blackshaw 2010, p.21). Bourdieu (1977) argued that 'doxa' contained definitional arbitrariness, that is, in terms of who decides what it means and influences how it is understood and accepted. People not only take on board the world as it is, but in doing so they reproduce it, albeit in their own way. For Bourdieu (1977), the 'established order' is replete with classifications, which suit its purpose; acceptance of this 'order' manifests in self-evidence. Therefore, according to Bourdieu (1977), the status quo is misrecognised as the only possible way, lacking challenge or alternatives. There is a 'political function' in such classification; by maintaining the status quo, people's place in the social world remains docile.

Care's particular 'doxicity' relates to a universal idea of warm feelings of support, yet it obscures levels of complexity and normative framings of how care manifests in practice: who does it, who deserves it, how governmental interventions shape ways of being in school populations (Foucault 2010). As such should someone stand outside of the norm, they risk experiencing non-care as a consequence. Tronto (2013) reflected on her earlier work (1993), which advocated that care ought to be central to democracy: "the world would look very different if we put care at the centre of our political lives" (Tronto 2013, p.x see also Moss 2014a). Tronto returned to the concept of care again in 2015, demonstrating there is still a long road to travel to ensure care is a central factor in decision-making, beyond the doxic rhetoric.

Domains of care

Clarity on what care means is therefore crucial for professional life, in order to practice ethically, and with regard to meeting need. Figure 2.1 offers a delineation of care as personal, public and private; each domain contains a 'duty' to care, but in different ways.

Care as a value is most associated with '*the personal*', that is the intimacy of the family, aligning with love and obligation to nurture. However, it does include a legal duty towards children – referred to as 'parental responsibility' (PR) to make decisions about health, education and discipline (HM Government 2014). The arrow on Figure 2.1 connotes the increasing privatisation of public services in

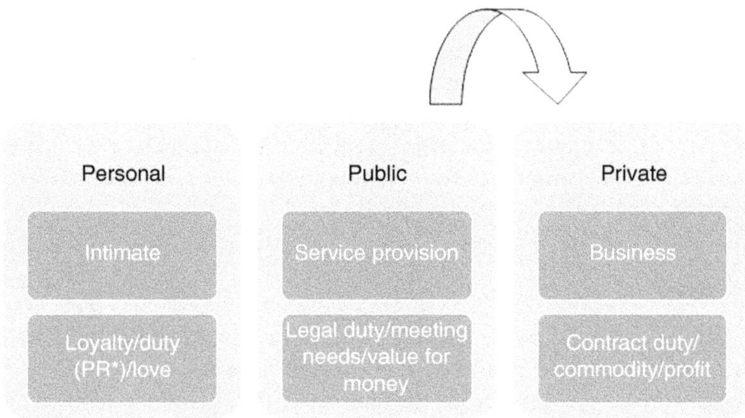

Figure 2.1 Multiple domains of care

England, such as contracting out elements of the National Health Service, transfer of social care duties from Social Services to private 'for profit' agencies, and in education the increasing use of Academies and Free Schools. This process releases some of the 'duty' from public services; care then manifests as a commodity in '*the private*' sector, with profit influencing the calculation of how care might be packaged and available to those in need.

Care is something services want to promote. McCuaig (2011, p.3), drawing from Goldstein (2002), identifies care as "a marketing slogan", seen where some school websites proclaim 'we care', emphasising commitment to inclusion, but without explication of how the claims to care within school ontology manifest, again requiring deconstruction. Moss (2014b, p.423, original emphasis) states categorically that "care is *not* a commodity" nor "only an activity" "done to" the "needy". Again connecting with Tronto's definitions above of '*actions and disposition*', Moss (2014b, p.423) suggests that care is a relational ethic with a mix of "doing and dispositions", where people recognise and respect "otherness and a rejection of calculative and rational thinking".

For the discussion here it is the 'public' domain of education which is of most concern. The 'Duty of Care' within public services in the UK is a requirement. In part it relates to *doing* health and safety, child protection and safeguarding, that is, policy-driven activity to keep young people safe. It relates to the responsibility charged when *in loco parentis*; "[a] court held, in 1893, that 'the schoolmaster is bound to take such care of his [sic] pupils as a careful father would'" (NUT 2012, p.3). In 1962 the role was professionalised, although the parental emphasis remained:

> Teachers' professionalism was recognised by the courts in 1962, where the 'standard of care' expected [required] the responsible mental qualities of a prudent parent in the circumstances of school, rather than home life. The current standard of care expected of a teacher is that of a reasonable person

> in the circumstances of a class teacher. It has been recognised that a teacher's duty of care to individual pupils is influenced by, for example, the subject or activity being taught, the age of the children, the available resources and the size of the class. (NUT 2012, p.3)

The latter part of this quotation identifies practicalities that impact the role and the importance of available resources for *doing* care; however it is harder to legislate for *disposition*.

States of care

School staff, as the responsible 'parent', generally sit within Figure 2.1's public domain; however, there is a need for a further layer of conceptualisation to explore *dispositions* situated within the 'intimate', 'duty' and 'relational' *states of care* (see Figure 2.2, and Hughes forthcoming). Such is the complexity of care that these *states of care* are porous where borders might be crossed in particular moments and contexts.

The notion of 'intimate' care in Figure 2.2 aligns with the 'personal' domain in Figure 2.1, with a greater expectation of unconditional emotional connection, garnering reciprocity, albeit not necessarily immediately because of the trusting nature of familial relationships. In terms of 'duty' Held (2006) suggests that "taking care of someone" can be viewed as work with an "expenditure of energy":

> the work of taking care of someone is not the same as caring for them in the sense of having warm feelings for them [intimate]. But whether certain feelings must accompany the labor of care is more in doubt [duty]. (p.30)

Held's position encapsulates the complexity of crossing from the personal/intimate to public/duty when operationalising care; where care is an action, feelings might not be present. In essence, the legislative 'duty of care' could be described as 'ought to', differing from the 'warmth' associated with natural caring drawn from 'want to' care. For Noddings (2013), natural care is where,

> [w]e just do care; no ethical effort is required. … This impulse arises naturally, at least occasionally, in the absence of pathology. We cannot demand that one have this impulse, but we shrink from the one who never has it. (p.81)

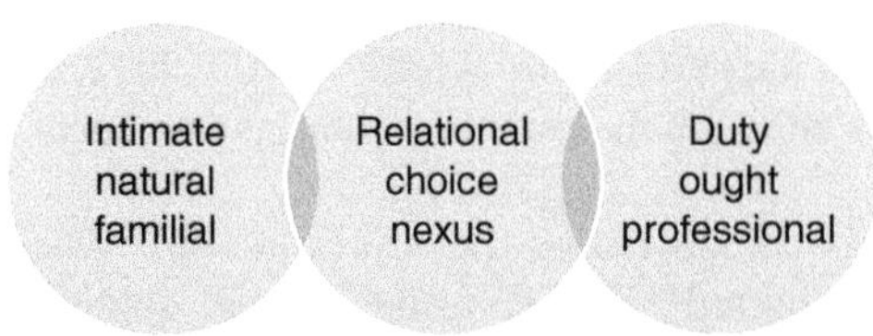

Figure 2.2 The caring nexus: multiple states of care

In a professional capacity 'ought' can be strained if the 'initial impulse' does not come naturally. Noddings (2013) suggests that it can be necessary to "fetch [care] out of recalcitrant slumber when it fails to awake spontaneously" (p.84). Ethical caring relates to the 'ought to'/'have to' duty when it may feel difficult to 'want to' care (Noddings 2013). Brannen and Heptinstall (2003) suggest that "care can be divorced from caring activity … conceptualise[d] [as] the practical activity of 'caring for' someone" as in a task.

Having a 'duty' ensures that, particularly where relationships fracture, a form of care is still in place, at best more than the minimal '*doing*' of policy provision. O'Connor (2008, p.117) examines similar terrain when exploring teacher roles, combining 'professional identity' and 'philosophical or humanistic beliefs about the teaching role'. For O'Connor (2008) "caring is primarily defined as those emotions, actions and reflections that result from a teacher's desire to motivate, help or inspire their students" (p.117). Indeed, O'Connor (2008, p.118) cites Hargreaves (1998, p.835) who in turn, draws from Palmer (1998) to suggest that "teaching is 'charged with positive emotion' and takes place at the intersection of personal and public life". O'Connor notes that Hargreaves (1994, p.175) ascribed to teaching "human nurturance, *connectedness*, warmth and love" (added emphasis). This is where linguistic boundary crossing takes place encompassing the 'natural', which is part of personal *disposition*/identity crossed with care as required; as 'ought to' the *doing*/professional duty.

Relational-care: the nexus between intimacy and duty

When considering the construction of care it is important to acknowledge intent, motivation, judgement, and in whose interests care is offered, what form it takes and what benefits are derived, for the young person, the staff, the school, government and so forth. In the spaces between personal/intimate with public/duty, the ideal 'state' would be '*relational-care*', which meets needs and is negotiated, contextual, socially just and reciprocal and thus ingrained with care. It would be *before* intimate, to delineate inappropriate crossing of professional boundaries, but *beyond* the baseline of duty, which just fulfils legal requirements, exacerbating conditions for disconnection. It is here that the dispositions of the significant adult facilitate working with young people to negotiate how to perform care in their interests.

The darker side of care

However, it is in such spaces that care can be precarious, misrecognised, manifesting as (non-)care.

Figure 2.3 offers two additional states, where the extremes of care fall into Barnes's (2012) 'dark side'. On the edges of intimacy, *overbearing care* can be guiltinduced coercion; *if you cared you would…*, something Petrie (2003) acknowledges as an attempt to secure conformity. Additionally it could manifest as relentless care as 'worry' (van Manen 2002), forcefully doing 'for' someone and undermining independence.

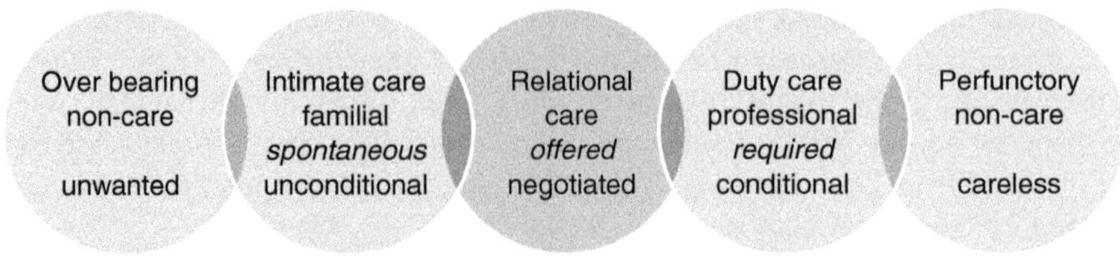

Figure 2.3 The caring nexus: including states of (non-)care

Brannen and Moss (2003, p.12) draw on Foucauldian notions of power as "all pervasive" within relations: "it both surrounds people and is wielded by them – we may be subjected to power, but we also exercise power". They choose a quotation to explicate this point:

> in human relations, whatever they are – whether it be a question of communicating verbally … or a question of a love relationship, an institutional or economic relationship – power is always present: I mean the relationship in which one wishes to direct the behaviour of another. (Foucault 1987, cited in Brannen and Moss 2003, p.12).

This alludes to the powerful forced nature of care. In Figure 2.3 the extreme 'state' of duty suggests care as '*perfunctory*', enforced application without feeling, not reaching 'good enough care' (Hollway 2006). It serves the legal obligation but in a manner that suggests no underpinnings of connection and no recognition of the person, their context or needs. Rather, it can be *careless* or dismissive. Vic, at a Northwick service, focused on young people's rights, and was commissioned to undertake research with the most persistent non-attendees at school. A teacher approached Vic with an illustration of dismissal by senior staff:

"So and so 50% attendance – *waste of space don't bother* – like that"
"So she [the teacher] says to me 'and that's what I am up against in this school' – 'that's [school name]' – how do I challenge that? *I tell you if I do challenge it I'll be 'sent to Coventry' [i.e. ostracised] in the staff room because I am in a minority* and *the head of year is very powerful*."

This illustrates the problematising of young people, but also exposes a further issue, of fracturing what Bourdieu (1977) refers to as habitus:

> the habitus is the product of the work of inculcation and appropriation necessary in order for those products of collective history, the objective structures (e.g. of language and economy, etc.) to succeed in reproducing themselves more or less completely, in the form of durable dispositions, in … individuals. (p.85)

Such *dispositions* guide ways of being and in doing so impact practice, as discussed by Tronto and Moss above. The example demonstrates ruptures in professional habitus within the school staff team, where values and *dispositions* clash, the staff member's desire to care in order to champion young people resulting in feeling their own sense of disconnection and alienation from seemingly non-caring colleagues.

Careless talk costs care

The dismissal of young people continues when messages are inculcated as 'injurious speech', a construct developed by Butler (1997) to understand which language 'wounds'. In this regard, Butler was influenced by Foucault's (1982, p.208) interest in how "human beings are made subjects". 'Injurious speech' illuminates long lasting imprints when messages offered by those in authority serve to subjugate influencing self-perception. Reay (2010, p.278) argues that "school remains the primary arena for engaging and dealing with a whole array of differences. Schools then are crucial sites of identity work and identity making." Certainly participants from Northwick were uncaringly positioned as non-normative by teacher 'experts' and their less than careful messages:

TONY [ADULT STUDENT]: "You won't make it in the army you know you'll end up in prison."

LUIS [IN NEET PROVISION]: "A few teachers said I'd just be laying in a dustbin when I got older and that's about it."

The adult students and the young people interviewed in the Northwick study could articulate experiences of non-care far more easily than they could articulate care. However, the four adult students' schooled self-perception of failed non-learner was interrupted via relational-care generated by staff at university, providing a 'turnaround' of their identity: to learner.

Constraining care – the reification of statistics

It is important to understand what might lead educational practitioners to come to the point where they dismiss the very young people they are employed to educate and to care about. There is no one sole reason; some include, for example, interpersonal differences, class, gender, ethnicity, social background and more. However, increasingly structural influences derived from neoliberalism impact the performance of care. Ball (2013) refers to 'performativity' concerning accountability, evaluation and competition between schools, problematising and stigmatising those who do not meet Ofsted (Office for Standards in Education, Children's Services and Skills in the UK) standards. The Head of a Pupil Referral Unit (PRU) in Northwick reflected that,

> I think the league tables haven't done us any favours ... because it focused mainstream schools on attainment and progress and meeting targets ... when I started off there used to be huge emphasis, certainly in inner cities, on

> pastoral care and welfare, knowing the parent and knowing the child and I don't think particularly the attainment was an issue because, like we get now with things like foundation learning programme and functional skills, there were those different pathways for kids to follow and they were thought as good as qualifications or accreditations as your routes through O' Levels and GCSE.

The Head draws out key issues, most importantly for this study, the emphasis on 'knowing' and understanding life-worlds, contextually shaping caring practice, conjoining *doing* with *disposition*. The quote highlights how status is conferred by attainment rather than including space for young people with differing abilities. The current instrumental target-driven approach to education has utilitarian consequences for the *few* whose efforts will not enhance the school league tables, and who may be sacrificed to exclusion units, in favour of the success of the *many* pupils who achieve and 'survive' the narrowly configured system. Certainly the Head of the PRU stated that the majority of their students had mild learning difficulties and could not keep pace with the speed of the national curriculum delivery. Reay (2012) concurs, reflecting that,

> what we need are totally different ways of envisioning education. Part of that process requires a revaluing of vocational and working class knowledges and a broadening out of what constitutes educational success beyond the narrowly academic. Much also needs to be done inside schools and classrooms to change the culture from the stressful, task-driven, target-led, overly competitive environments they currently are. (p.3)

Cooper et al (2015) join Reay in arguing for socially-just, radical alternatives to ways of working with young people. In the following quotation Andrew [a former deputy Headteacher, now 'Turnaround' co-ordinator in Northwick] offers an insight into how the concentration on numbers impacts teachers:

> I mean so yes in the primary school the smaller it is that constant relationship with the class teacher it is easier to do that, but there is so much pressure nowadays from the government on SATS that a lot of the opportunities are negated because teachers, and I say this from speaking to a lot, many in fact, are very disillusioned, disenchanted by the bureaucratic nightmare of self-assessment, monitoring frameworks coming in.

Andrew's point was borne out by most research participants, that primary school was more likely to radiate care through its smaller environment, and time to build connectedness with the teacher. The size and transience of secondary school impacts caring relational encounters detrimentally. Crucially, Andrew explicates teachers' own levels of disconnection with the education system, mirroring the young people they work with and constraining ability to care (see also Hughes et al [2014] relating to mirrored oppression in youth work practice).

The time it takes for teachers to undertake the bureaucratic aspects of their work replaces time spent in creating relationships, leading to some participants in Northwick feeling reified as statistics:

GILL: "Do you think they [teachers] cared about your learning?"
RENEE [AN ADULT STUDENT]: "No I think they cared about getting the results they needed."

This experience was echoed by younger participants, contributing to the 'exam factory' analogy connoted by Coffield and Williamson (2011). Levels of care and thus connectedness are detrimentally impacted when time-poor staff privilege statistics over students, resulting, for many, in disconnectedness, which impedes the neoliberal aim to produce fiscally active citizens. However, if care creates connectedness and in turn active citizens who contribute to the economy, then in appropriating the language of managerialism, ironically care could be argued to be *cost effective* if connectedness facilitates transition to the marketplace (see also Hughes 2015).

Who cares and how would you know? A contextualised lens of caring judgement

The opening question suggests the complexity of understanding care and how it manifests. This chapter has taken account of its subjectivity and the borders of non/care, which complicate its definition, and introduced a range of conceptualisations to peel back the layers in order to examine the nuances of the construct. This chapter argues for a non-judgemental space in which practitioners can examine care personally and professionally, to ensure they can deconstruct, think, reflect, discuss and debate the role of care in practice. Todres (2008) asserts

> that there is a 'goodness of fit' between embodied relational understanding – the practice of care that is judgment-based – and an emphasis in qualitative research that can better serve such concerns. (p.1566)

Research-infused learning and teaching could provide practitioners the opportunity for reflection and reflexivity across professions. Brannen and Moss (2003) explicate this within education as

> a pressing need for all involved in children's care to have more space – to read, think, discuss, write, problematize and evaluate – about theory, practice, research, the world we live in today and the past from which it comes. We say this not as academics situated above the forces and pressures in an increasingly marketized research arena. Only by valuing and creating such space can rich and provoking processes of rethinking occur – and without such rethinking, restructuring is simply a superficial exercise. (p.199)

The ontological misrecognition which can manifest between colleagues, fracturing habitus within the professional field, might be alleviated with collegiate engagement and open discussion on what care means to each person, emphasising a central tenet of this chapter that care must be negotiated within context, to perform connection to education. Trout (2012, p.3) reflects that,

> teachers … play a pivotal role in how relationships develop between members of classroom and school communities. Surprisingly, however, in the halls of teacher education programs and the traditional checklists that evaluate practicing teachers, little emphasis is placed on helping beginning teachers learn how to cultivate relationships for pedagogical purposes … we know little about how beginning teachers learn to build relationships and facilitate student learning. (p.3)

Trout is not alone with this claim. Goldstein and Lake's (2000) module aimed to facilitate an exploration of care in practice beyond an assumption that; teaching is synonymous with care.

There are advantages to the possibilities afforded by employing the doxicity of care. It could be viewed as an opportunity to counter universal interventions through negotiating what care should look like without strict boundaries; however, it is also important that the same doxicity does not lead to oppressive practices. Practitioners must be critically conscious of its amorphous nature, to ensure that it is situated within an ethical framework. Tronto suggested that care should form a 'habit of mind' (1993), and more recently, Tronto (2015, 2013) promoted 'care-with', suggesting intent towards an equal engagement, placing care at the centre of democracy. Barnes (2012) offers 'a lens of care', while Sevenhuijsen (1998, p.15), aligns care and justice, positing that "judging with care" draws on "their [practitioners'] expertise and moral considerations", beyond a "fixed caring identity", eschewing labels by promoting "different but equal" treatment, dictated by the context of the situation. These various conceptualisations foreground *thinking with care* to advance the possibilities of developing a '*contextual lens of caring judgement*', positioning care centrally to the culture of any society, infusing relational encounters, decision-making, policy development and operationalising interventions, such as connectedness. This lens encompasses the ideal state of *relational-care*, imbued with context, social justice, reciprocity, negotiation and caring judgement, to identify and respond to needs articulated by the caree, and not to the detriment of the practitioner.

Fine (2007, p.18) calls for "a new awareness [of] care as a system of values and consciousness … [creating] … enormous potential to re-shape the way modern societies are organised". This conjoined with Tronto's caring democracy (2015, 2013) and Reay's call to change education could powerfully provide the conditions to facilitate a new landscape of ontological practices, thus challenging the neoliberal education system and its shaping of 'products' for the marketplace, and embracing relational-care in ways that meet the needs of young people and practitioners to forge connectedness to education.

References

Ball, S.J. (2013) *Foucault, power, and education*. London, Routledge.

Banks, S., Armstrong, A., Carter, K., Graham, H., Hayward, P., Henry, A., Holland, T., Holmes, C., Lee, A., McNulty, A., Moore, N., Nayling, N., Stokoe, A. and Strachan, A. (2013) Everyday ethics in community-based participatory research. *Contemporary Social Science* 8, pp.263–77.

Barber, B.K. and Schluterman, J.M. (2008) Connectedness in the lives of children and adolescents: a call for greater conceptual clarity. *Journal of Adolescent Health* 43, pp.209–16.

Barnes, M. (1997) *Care, communities and citizens*. London, Longman.

Barnes, M. (2012) *Care in everyday life: an ethic of practice*. Bristol, Policy Press.

Barker, D. K. (2014) Beyond women and economics: rereading 'women's work'. *Signs 40*, pp. 2189–2209.

Blackshaw, T. (2010) *Key concepts in community studies*. London, Sage.

Booth, T. and Ainscow, M. (eds.) (1998) *From them to us: an international study of inclusion in education*. London, Routledge.

Bourdieu, P. (1977) *Outline of a theory of practice*. Cambridge, Cambridge University Press.

Brannen, J. and Moss, P. (2003) *Rethinking children's care*. Milton Keynes, Open University Press.

Brannen J. and Heptinstall, E. (2003) Concepts of care and children's contribution to family life. In Brannen, J. and Moss, P. (eds.) *Rethinking children's care*. Buckingham, Open University Press, pp.185–197.

Butler, J. (1997) *Excitable speech: a politics of the performative*. Abingdon, Routledge.

Chakelian, A. (2015) Budget 2015. What welfare changes did George Osborne announce, and what do they mean? *The New Statesman*. 8 July, 2015. Available at: www.newstatesman.com/staggers/2015/07/budget-2015-what-welfare-changes-did-george-osborne-announce-and-what-do-they-mean

Chapman. R.L., Buckley. L, Sheehan. M. and Shochet. I. (2013) School-based programs for increasing connectedness and reducing risk behaviour: a systematic review. *Educational Psychology Review* 25, pp.95–114.

Coffield, F. and Williamson, B (2011) *From exam factories to communities of discovery*. London, Institute of Education

Cooper, C.E., Gormally. S. and Hughes G. (eds.) (2015) *Socially-just radical alternatives for education and youth work practice: re-imagining ways of working with young people*. Basingstoke, Palgrave.

Dunne, L. (2008) Discourses of inclusion: a critique. Paper presented at British Educational Research Association Annual Conference, 3–6 September, Heriot-Watt University, UK.

Fernandez, A.L., and Gabe, T. (2009) *Disconnected youth: a look at 16- to 24- year olds who are not working or in school*. Darby, DIANE.

Fine, M.D. (2007) *A caring society? Care and the dilemmas of human service in the twenty-first century*. Basingstoke, Palgrave.

Fisher, B. and Tronto, J. (1990) Towards a feminist theory of caring. In Abel, E. and Nelson, M. (eds.) *Circles of care: work and identity in women's lives*. Albany, State University of New York Press, pp.35–62.

Foucault, M. (1982) The subject and power. In Dreyfus, H. and Rabinow, P. (eds.) *Michel Foucault: beyond structuralism and hermeneutics*. Chicago, University of Chicago Press, pp.208–26.

Foucault, M. (1988) Truth, power, self: an interview with Michel Foucault, October 25, 1982. In Martin, L., Gutman, H. and Hutton, P. (eds.) *Technologies of the self: a seminar with Michel Foucault*. Amherst, University of Massachusetts Press, pp.9–15

Foucault, M. (2010) *The birth of biopolitics: lectures at the Collège de France, 1978–1979*. Basingstoke, Palgrave.

Gilligan, C. (1982) *In a different voice: psychological theory and women's development.* Cambridge, Harvard University Press.
Goldstein, L.S. and Lake, V.E. (2000) 'Love, love, and more love for children': exploring pre-service teachers' understandings of caring. *Teaching and Teacher Education* 16, pp.861–72.
Goldstein, L. (2002) *Reclaiming caring in teaching and teacher education.* New York, Peter Lang.
Hansen, P. and Mulholland, J. A. (2005) Caring and elementary teaching the concerns of male beginning teachers. *Journal of Teacher Education*, 56(2), pp.119–31.
Hargreaves, A. (1994) *Changing teachers, changing times: Teachers' work and culture in the postmodern age.* London: Cassell.
Hargreaves, A. (1998). The emotional practice of teaching. *Teaching and Teacher Education,* 14(8), pp.835–854.
Hark, S., and Villa, I.P. (2011) 'Confessing a passionate state': Interview with Judith Butler. *Feminist Studies* 29(2), pp.196–205
Held, V. (2006) *The ethics of care: personal, political, and global.* Oxford. Oxford University Press
HM Government (2014) *Parental rights and responsibilities.* Available at: www.gov.uk/parental-rights-responsibilities/what-is-parental-responsibility
Hollway, W. (2006) *The capacity to care: gender and ethical subjectivity.* London, Routledge.
Hughes, G. (forthcoming) *Who cares and how would you know? (Dis)connectedness, education and care.* Unpublished PhD thesis, University of Hull.
Hughes, G. (2015) 'Imagining otherwise' or tinkering with the system? In Cooper, C.E., Gormally. S. and Hughes G. (eds.). *Socially-just radical alternatives for education and youth work practice: re-imagining ways of working with young people.* Basingstoke, Palgrave, pp.220–44.
Hughes, G., Cooper, C., Gormally, S. and Rippingale, J. (2014) The state of youth work in austerity England: reclaiming the ability to 'care'. *Youth and Policy* 113, pp.1–14.
McCuaig, L. (2011) Dangerous Carers: Pastoral power and the caring teacher of contemporary Australian schooling. *Educational Theory and Philosophy* 44(8), pp.862–877.
Moss, P. (2014a) The social protection floor: what place for care? *Global Social Policy* 14, pp.422–31.
Moss, P. (2014b) *Transformative change and real utopias in early childhood education.* London, Routledge.
Noddings, N. (1984) *Caring: a feminine approach to ethics and moral education.* Berkeley, University of California Press.
Noddings, N. (2002) *Starting at home: caring and social policy.* Berkeley, University of California Press.
Noddings. N. (2003) 'Is teaching a practice?' *Journal of Philosophy of Education* 37, pp.241–51.
Noddings, N. (2013) *Caring: a relational approach to ethics and moral education.* Berkeley, University of California Press.
NUT [National Union of Teachers] (2012) *NUT Notes 2012-2013.* Available at: www.teachers.org.uk/files/the-law-and-you–8251-.pdf
O'Connor, K.E. (2008) 'You choose to care': teachers, emotions and professional identity. *Teaching and teacher education* 24, pp.117–26.
Petrie, P. (2003) Social pedagogy: an historical account of care and education as social control. In Brannen, J. and Moss, P. (2003) *Rethinking children's care.* Milton Keynes, Open University Press, pp.61–79.
Philip, C., Rogers, C. and Weller, S. (2013) Understanding care and thinking with care. In Rogers, C. and Weller, S. (eds.) *Critical approaches to care: understanding caring relations, identities and cultures.* London, Routledge, pp.1–12.
Reay, D. (2010) Identity-making in schools and classrooms. In Wetherell, M. and Mohanty, C.T. (2010) *The SAGE handbook of identities.* London, Sage, pp.277–95.

Reay, D. (2012) What would a socially just education system look like? Available at: http://classonline.org.uk/docs/2012_Diane_Reay_-_a_socially_just_education_system.pdf
Rogers, D.L. and Webb, J. (1991) The ethic of caring in teacher education. *Journal of Teacher Education* 42, pp.173–81.
Rogers, C. and Weller, S. (eds.) (2013) *Critical approaches to care: understanding caring relations, identities and cultures*. London, Routledge.
Ruddick, S. (1996) Injustice in families: assault and domination. In Snow, N. (ed.) *In the company of others*. London, Rowman and Littlefield, pp.65–90.
Sevenhuijsen, S. (1998) *Citizenship and the ethics of care: feminist considerations on justice, morality, and politics*. London, Routledge.
Todres, L. (2008) Being with that: the relevance of embodied understanding for practice. *Qualitative Health Research* 18, pp.1566–73.
Tronto, J. (1993) *Moral boundaries: a political argument for an ethic of care*. London. Routledge
Tronto, J. (2013) *Caring democracy: markets, equality, and justice*. New York, New York University Press.
Tronto, J. (2015) *Who Cares? How to Reshape a Democratic Politics*. New York, Cornell University Press.
Trout, M. (2012) *Making the moment matter: care theory for teacher learning*. New York, Springer.
van Manen, M. (2002) Care-as-worry, or 'don't worry be happy'. *Qualitative Health Research* 12, pp.264–80.
van Manen, M. (1990) *Researching lived experience: human science for an action sensitive pedagogy*. Albany, State University of New York Press.
Vogt, F. (2002). A caring teacher: explorations into primary school teachers' professional identity and ethic of care. Gender and Education, 14(3), 251–264.
Wingspread Conference (2004) Wingspread declaration on school connections. *Journal of School Health* 74, pp.233–34.

3 Theories underpinning kinship care

Rebecca C. Pratchett and Paul Rees

Introduction

Every year in England and Wales around 30,000 children and young people are removed from their parental home and placed in alternative forms of care (DfE 2015, Statistics for Wales 2015). Whilst historically such children would have been placed for adoption, local authorities increasingly seek to place children with extended family in kinship care. Alternative family-based care is far from a new concept and is evident in a diverse range of cultures. However, a rise in 'formal' kinship placements, defined as those that are known to and regulated by the State, warrants a fresh exploration of the theoretical basis of kinship care.

In this chapter we bring together key theories from a range of disciplines in order to understand the development of kinship care in social work. We introduce social work in the context of structural-functionalism in order to explain the development of the field. We discuss the relationship between the family and the State using the concept of dynamic equilibrium. In doing so we recognise the changing nature of social work, both in its structural composition and its functions. We identify patterns of change in practice which represent the reactive nature of social work and its developing functions in a changing society. Crucially, we discuss its role in the preservation of the family and propose that social work evolved as a response to the breakdown of the nuclear family.

The chapter explores the biological, sociological and psychological theories of kinship organisation. We discuss the genetic basis of altruism and its limitations in explaining the altruistic nature of kinship care. Through a critique of the biological explanations of kinship, this chapter moves towards a more sociological understanding of the family unit. In particular we propose, via the stance of social identity theory, that a genetic basis for kinship care is not necessarily a prerequisite and that the altruistic actions of carers may be explained by group selection. In the final section we collate theories of family and social work, theorising the nature of their interaction. Using Bronfenbrenner's (1992) ecological systems theory, we seek to demonstrate how this interaction has developed over time to create the contemporary social work model. We propose the framework for kinship care in social work and define the potential benefits of its adoption. In conclusion we hypothesise that kinship care has the potential to be mutually beneficial to the

State and to young people requiring alternative care provision. With appropriate application, kinship care has the potential to reduce the 'looked after' population, promote resilience within the family unit and break the intergenerational cycles of care dependency.

Kinship care in context

The question of how to respond to the needs of children who are unable to live with their birth parents presents dilemmas for every society and many have been reluctant to take on responsibility for providing the necessary care (Rees 2010). Even in industrialised nations such as the United Kingdom (UK) there is a relatively short history of State intervention and provision of care for children. The concept and practice of familial wardship which was embedded in, for example, the *Poor Law Amendment Act* (1834) meant that grandparents and other close relatives were held legally responsible for providing care to their grandchildren should the child's parents be unable to do so. Despite the immense social, economic and political change that occurred between the time of the Act and the mid-20th century the primary importance of familial responsibility held firm.

It was not until the *Children Act* (1989) that the 'State' unequivocally assumed responsibility for these children by requiring local authorities to provide accommodation when no person had parental responsibility, or when the person(s) with parental responsibility was unable to provide, for whatever reason, suitable accommodation or care (*Children Act 1989*, Section 20). It is highly significant that these children are referred to in the *Children Act* (1989) as 'looked after', confirming the State's acceptance of the role of 'care provider'. This role has grown and the 'looked after' population in the UK is now close to 90,000 (DfE 2014, Waugh 2014, Scottish Government 2015, Welsh Government, 2015). Although some forms of 'State' care such as foster placements and residential homes have been numerically more popular, formal kinship care has also been a favoured option in a small proportion of cases. The social, health and educational outcomes of 'looked after children' are varied with some performing well in many domains (Rees 2013), but there is also some evidence to suggest that 'looked after children' in kinship care perform relatively well (Winokur et al 2008).

Recently there has been an increase in the number and proportion of children being placed with relatives upon removal from the nuclear family with around 12% of all 'looked after children' in England now living in formal kinship care (Selwyn 2009). The underlying reason for this growth is unclear although major care scandals, poor care leaver outcomes, a shortage of mainstream foster carers and decreasing local authority budgets have all been proposed as influencing factors (Biehal et al 1995, Broad 1998, National Foster Care Association 1997). Interestingly, a similar trend is thought to have occurred in respect of 'informal' kinship care: full-time care provided by kin which has been privately arranged without the State's involvement. Using national census returns Selwyn and Nandy have estimated the total number of children in the UK being brought up by relative to be around 173,000 which provides an indication of how widespread the practice has become. Given the fact that the State is often unaware of informal kinship care and has no monitoring role it has no means of knowing why the practice is so prevalent. It would seem reasonable to hypothesise that the media's portrayal of some social workers as, at times, unduly interventionist and inept may encourage guarded behaviour by family networks. Similarly, an increase in the necessity for geographically mobility on the part of bread-winners, who also wish to maintain social and educational continuity for their children, would seem to be a plausible explanation for some of the growth in private kinship arrangements. Further research which is informed by theory is needed if a clearer picture is to be established.

Here we discuss the assertion that rather than solely being a reaction to such factors, the increase in formal kinship care may also represents an ideological shift towards a strength-based model of care through the fortification of the family unit. Given the historic role of family, it is important to understand the theoretical basis of social care of children in modern society, and the role family continues to play. It is imperative that we understand how the two interact to create the modern social work model of alternative care for children. This chapter will examine the sociological and biological basis of kinship care to provide novel insights into the theoretical underpinnings of modern social work practice and the role of the family within it.

Structural-functional theory and the evolution of social care work

To begin to understand the theoretical framework of kinship care, it is important to understand the broader social context within which kinship, and even social work as a societal institution, exist. In order to do so, this section will examine social work from a wider, sociological perspective with a particular focus on structural-functionalism. This section aims to assess the role of social work in both traditional Western society and in a modern day context.

Society as a construct is widely debated (Giddens et al 2003), although it can be understood through the perspective of structural-functionalism. This theory posits that society is a collection of social systems that interact to maintain order within society. Based upon Durkheim's functionalist approach (Pope 1975), these social systems are groups and institutions of varying size, and share cultural values and norms. They are cohesive in allowing members to identify their own place in society and to relate to others who they deem to be their peers. Structural-functionalism differs from Durkheim's functionalist theory in that the individual is said to be irrelevant and that institutions function to serve the collective rather than benefit the individual. Within a structural-functionalist perspective, large institutions such as religions or government are important in sharing and enforcing moral and legal norms. Education also becomes an important structure for socialisation and obedience.

The most common analogy for structural-functionalism, made popular by Herbert Spencer (Macionis and Plummer 2009) amongst others, compares the structure of society to the human body. Within this analogy, institutions perform functions in the same way that organs perform functions within the body. The failure or dysfunction of any of the institutions will cause widespread dysfunction amongst others just as the failure of an organ does. Therefore, the overall goal of society is to create an equilibrium where social processes maintain balance and prevent the failure of any institution. Equilibrium is an important concept in functionalist thought. Functionalists such as Durkheim recognised that society is in a constant state of flux and that as change occurs, society adapts to maintain equilibrium. Parsons described this constant fluctuation as dynamic equilibrium (Parsons 2013). For example a decrease in employment may lead to higher rates of criminal activity. In functionalism, this adaptation is an implicit action which occurs without any conscious effort or acknowledgement. By contrast, structural-functionalists recognise two types of function. Robert K. Merton (1957) defined these as manifest and latent functions. Manifest functions are the deliberate or anticipated function of a structure, whilst latent functions are the unintended or unanticipated functions. The cycle to work scheme, introduced in the UK in 1999, is a good example to highlight the differences between latent and manifest functions. Cycle to work enables people to buy a bicycle tax-free via their employer. The manifest functions of the scheme were two-fold: to create a healthier workforce who would require lower health spending and less time off work, and to reduce the environmental impact of the work commute. The scheme also has latent functions in that inner congestion was reduced and the cycling sector has a financial boost. However, there were some negative consequences, known as dysfunctions, such as an increase in cycling-related death and injury and an increase in bicycle theft (Home Office 2011).

For structural-functionalists, the family is an important institution, which represents the building block of society. Referring back to the biological model, the family, or more specifically the nuclear family, performs specific societal functions. Specifically, Murdock (1949) noted four functions fulfilled by the family: catering to sexual needs and limiting sexual access of other members of society; reproductive functions; economic functions with a focus on the division of labour within a family unit to allow prosperity; and the education of the young within the family, also known as 'socialisation'. Each of these functions have an important role in the production and maintenance of citizens who can then become active participants in society. By socialising the young, cultural values and norms can be transmitted to a new generation to prevent deviance and maintain equilibrium.

With the erosion of the reality of the single breadwinner family, the nuclear model of family no longer fulfils the four functions outlined in structural-functionalist thought. In families where both parents work, the division of labour has evolved and many of the more domestic duties are often outsourced to third parties. In some cases services such as childcare, traditionally the role of the mother in a nuclear family, are provided by a formal organisation. However, in around 40 per cent of families, this role is adopted by a family member such as a grandparent (Huskinson et al 2014). In such cases the division of labour is not simply between the patriarch and matriarch of the modern family, but may include other family members. A wider understanding of family to include extended kin provides a more representative approach in a modern context whilst encapsulating the four primary functions of family as identified by Murdock (1949).

Whilst the family unit is universally recognised, the composition varies throughout the world. Some societies focus on community-based child rearing as demonstrated by the African proverb, 'it takes a whole village to raise a child'. As such, structural-functionalism has suffered heavy criticism for its emphasis on the nuclear family. Certainly in a modern Western society where single-parenthood is increasingly common and family structure has shifted away from the nuclear model, it is important to consider the relevance of a model based on such a limited definition of family. However, rather than disregard the model, it could be adjusted to incorporate a more modern concept of family.

While family is recognised as a bedrock of society we must try and understand how social work has evolved to fulfil the functions that it does. Historically, social work may be understood more as an approach than a specific and identifiable structure. With its origins in the work of a variety of religious and charitable organisations, social work served to protect vulnerable children from the dysfunction that resulted from the breakdown of the nuclear family. Over the years it has become a more organised and substantial entity, although to identify it as a structure with its own functions would perhaps be misleading. Rather, social work seems to serve to reinforce the functions of other structures, in particular the family. Originally this was achieved through the child rescue model where a child was removed from a dysfunctional family and placed into a substitute nuclear family who would be able to fulfil the necessary functions and produce a stable citizen. Similarly, the shift in social work practice which emphasised family-based interventions rather than removing the child simply served to reinforce the structure and functions of the family. It could therefore be argued that social work is a sub-unit of the nuclear family and serves not only to support the functions of the family in society, but also to maintain its structure.

As the 'building block' of a societal system assembled from inter-reliant structures, the family is a critical institution in the structural-functionalist model. When considered in such a context, the relationship between family and State becomes clearer. Social work, as an institution in its own right, is a vehicle to ensure the stability of the family thus providing a stable bedrock for society to develop.

Theories of kinship

The previous section gave a broad overview of the social context within which social work functions as an institution. This section will examine theories of kinship through biological and sociological perspectives in order to understand the importance of kin relations. Through these theories, we will also consider some of the motivations that lead kinship carers to become the primary carer for young people in the care system. According to the *Oxford English Dictionary*

(OED), kinship can be understood by two definitions. The first, a more biological approach, defines kinship as "relationship by descent" (OED 2016b). This focuses on genetic lineage and therefore the scientifically quantifiable relationship between individuals. In a more anthropological approach, kinship is also defined as "the recognised ties of relationship by descent, marriage or ritual that form the basis of social organisation" (OED 2016c). Whilst it encompasses the genetic element of kinship, this broader definition understands kinship to be the social relations between individuals. These relations, whether formalised through religious rituals such as marriage or through social rituals, become the basis for the organisation of individuals into kinship groups. This may also have grown to include friends of the family who were more traditionally referred to as 'kith'. These differing definitions highlight the ontological conflict in understanding kinship and can be used to examine the proposed theories of kinship relations.

In some respects, kinship appears to conflict with scientific theories of species survival. According to theories of evolution, organisms within a species that have genetic mutations that prove advantageous to their survival will thrive and pass on these genetic benefits. Those who do not are more likely to perish and so fail to reproduce. It would therefore seem contradictory that altruistic traits would continue to be present in both humans and other species. Altruism is best understood as a "behaviour of an animal that benefits one or more others (typically of its own species), but which carries a cost for the individual concerned" (OED 2016a). In many cases, the cost of altruism is high and can include mortality. Velvet monkeys, for example, have been known to make alarm calls when under attack to warn others within their group of the danger. Whilst this is beneficial for the rest of the group, these cries may also attract the attention of the predator and make the individual more vulnerable to attack (Cheney and Seyfarth 1992). Similarly, the financial, lifestyle and time costs associated with becoming a kinship carer make the act of doing so an altruistic one.

How then has altruism survived natural selection? Evolutionary psychologists claim the answer lies in two theories: selfish gene theory and inclusive fitness. Richard Dawkins popularised the concept of the selfish gene in a 1976 book of the same name (Dawkins 1976). Building upon previous theories of evolution proposed by George C. Williams (1966), Dawkins argues that genes have the single purpose of replicating to ensure its continued existence. Whilst the use of the term 'selfish' ascribes a confounding level of anthropomorphism to the innate action of genes, the theory itself provides a solid foundation for the theory of inclusive fitness. This theory posits that our actions are genetically driven by the innate need to replicate. In order to ensure the survival of genes, an individual may sacrifice themselves in order to protect their offspring or other biological relation. This is arguably evidenced by matriphagy, the apparent sacrifice of the maternal spiders to provide nutrients to offspring in some species (Toyama 2001). Hamilton even proposed that inclusive fitness is a vehicle that promotes altruistic traits and in doing so created a mathematical formula to calculate whether the altruistic gene would pervade through a population (Hamilton 1964a, 1964b).

Some might be tempted to view inclusive fitness as an explanation for the motivations of kinship carers. Kin, as described in a biological sense, may have a genetic predisposition to behave altruistically in order to preserve the genetic lineage. This is summarised well by J.B.S. Haldane who once quipped: "I'd lay down my life for two brothers or eight cousins" (Lewin 1974). In doing so, Haldane humorously suggests that altruistic behaviour is directly proportional to the degree of genetic relatedness to another. Whilst the biological explanations of kinship present some interesting observations in the animal world, the extent to which they can be applied to human behaviour is, arguably, limited.

One of the main challenges to theories of inclusive fitness is the adoption of unrelated infants and children. Proponents of inclusive fitness argue that adoption of unrelated individuals is an 'off-track' expansion of kin-adoption that has historically prevailed (Barrett et al 2002, Buss 2015). By describing adoption as 'off-track' these authors seek to dismiss cases of adoption as outliers, or as a phenomenon created by social conditioning. This view is seemingly supported by evidence of adopted individuals being treated less preferentially than genetically related young (Daly and Wilson, 1980; Eberhard, 1975). Regardless, the adoption of an unrelated individual seems to counter the arguments of inclusive fitness as it removes financial and emotional resources from one's own genetic offspring. It seems counterintuitive that a kinship unit would expend resources on those with no genetic linkage to themselves. Wilson, historically a major proponent of kin selection, co-authored a paper which refuted inclusive fitness in favour of group selection (Nowak et al 2010). Here the formation of social groups was recognised as a more important factor in determining the altruistic behaviour of both humans and other species. This would certainly seem to provide a more satisfactory explanation of kith care and adoption although it does not fully address the notion of the latter, particularly where there seems to be no identifiable grouping from which the altruistic motivation arises. To understand this further, we must move away from the sociobiological explanations of kinship and turn our attention towards the fields of psychology and sociology.

Social identity theory is one of the most recognised theories to emerge from the field of social psychology. Introduced by Henri Tajfel in the 1970s, social identity theory focuses on how we identify with various groups in society and how we draw our self-esteem from the accomplishments of the groups to which we consider ourselves to belong. These groups are referred to as 'in-groups' and studies have demonstrated a tendency to behave preferentially toward our in-group when compared with 'out-groups'. In particular, Tajfel and colleagues found that even when assigned to a group arbitrarily, study participants allocated more points to their in-group than their out-group (Tajfel and Billig 1974, Tajfel et al 1971). It would seem therefore that we require little more than association with a certain group in order to behave favourably towards its members. It is through this that proponents of social identity theory posit that our social identity, as determined by the groups we associate with, can impact upon our behaviour.

Perhaps the most recognised study of social identity comes from the Stanford prison experiment conducted by Zimbardo and colleagues (Haney et al 1973). In this study a mock prison was created and twenty-four students were assigned either the role of prisoner or guard. This role allocation was entirely arbitrary and the participants were given no instruction on how to behave. Originally designed to last for a fortnight, the study was called off after only six days due to the emotional distress of the prisoners who were suffering at the hands of the guards. Despite being given no instruction or training on how to fulfil the role of a guard, the participants allocated to the role behaved in a similar way to real prison guards, using both physical and psychological punishments to maintain order and authority. This phenomenon was eloquently described by Abrams and Hogg (1990, p.297) who stated that "group behaviour arises from a shared sense of social category membership". In the Stanford prison experiment this was demonstrated by the guards who identified as a part of that group and so behaved as they believed a guard should.

Numerous studies have informed us that we will behave preferentially to a group to which we are assigned in a laboratory setting. What is interesting, however, is how this can be applied to enhance our understanding of natural group dynamics in everyday life. A single person is normally a part of numerous in-groups, whether it be a small group such as a sports team, or a large group such as a religion or nation. The different groups to which we are affiliated are ones with which we may share certain values or beliefs and to which we identify as being a part of. In the words of Emerson (1960), "the simplest statement that can be made about a nation is that it is a body of people who feel that they are a nation". Simply, in order to be a part of a group one must only feel a part of that group.

In the first section we identified the structural-functionalist perspective of the family as the base unit of society. Through this model, the concept of family has been reinforced for centuries. The biological models explored earlier in this section went some way to explaining how the kinship group came to form in nature, however kin selection could not sufficiently account for kith relations and adoption. By contrast, social identity theory may go some way to explain kith relations and their role in the social group. Social identity theory posits that a family group is defined by each individual's understanding of themselves within the social group. Therefore an individual who considers him/herself to be part of a family group may be so in spite of a lack of biological relatedness. This may also be indicative of an innate and implicit understanding of the relationship between nature and nurture. This may widen the traditional understanding of kin to encapsulate kith and go some way to explaining the motivations of kith carers.

Importantly, social identity theory provides a further explanation of the factors that motivate kinship carers to undertake caring duties despite negative consequences for their own wellbeing. In a study of informal kinship carers, researchers found that carers often described a sense of ownership and family obligation (Selwyn et al 2013, p.13). This sense of obligation may come from the internal identity of the carers who see it as part of the role they should play. This concept is supported by the number of female kinship carers. Traditionally caring roles were the responsibility of females within the nuclear family. Despite decades of social change leading to the sharing of such roles, females still account for more than 55 per cent of kinship carers in the UK (Nandy et al 2011). As such the figures would suggest that the roles in which we identify within the family unit may not have changed as much as we may wish to imagine. Equally, the function of the family unit as a protective factor seems to be engrained in our perceptions of kinship relations.

One of the major strengths of social identity theory is that it does not seek to over-ride the social, political and biological explanations of intergroup relations, but rather to complement theories in these areas. In doing so we can see how social identity theory enhances our understanding of group dynamics to supplement our comprehension of group selection. In the final section of this chapter, we will explore how social work and kinship relations interact to fulfil the functions of the family.

The interaction of social work and kinship

In the previous sections we examined the roles of both social work and family in society and explored the theoretical basis of kinship. In this section we will explore how social work, as we have come to understand it, interacts with families to form kinship care placements. In order to do so, we shall discuss ecological systems theory and how it is integrated into the modern day social work model.

Ecological systems theory was formulated by Urie Bronfenbrenner, an American psychologist working in the field of child development (Bronfenbrenner, 1979, 1992). This theory was founded on the view that children do not develop within a 'vacuum' but rather develop within the social, economic and political climates that surround them. It followed, therefore, that the external environments at the various levels of interaction would impact upon the child's development. Whereas previously child development had focussed on the immediate environment of the child, Bronfenbrenner introduced to the field a wider perspective within which developmental psychologists should work (Bronfenbrenner 1977).

The ecological model can be understood as five embedded structures resting inside one another "like a set of Russian dolls" (Bronfenbrenner 1994, p.39). This structure can be seen in Figure 3.1. The microsystem consists of the child's immediate environment and includes the physical, social and interpersonal interactions that the child faces on a day-to-day basis. Within the microsystem may be the immediate family relations, friends, and school interactions such as with peers and teachers. The mesosystem consists of the "linkages and processes between two or more settings containing the developing person" (Bronfenbrenner 1994, p.40). Essentially the mesosystem is a series of microsystems interacting with one another such as the parent and teacher. The next layer, the exosystem, contains the links and processes between a setting where the developing person has no active role and the individual's immediate context. An example would be a parent acquiring a new job which requires them to work longer hours and so reducing the amount of time the parent could spend at home with the developing person.

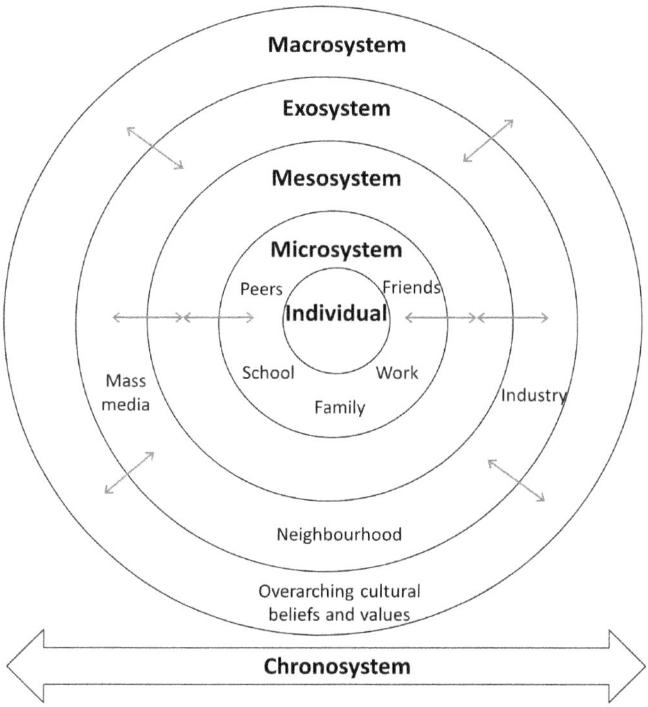

Figure 3.1 Brofenbrenner's ecological model of human development (adapted from Bronfenbrenner 1992, 1994)

The macrosystem refers to the cultural context within which the developing person lives. This can be, for example, ethnicity or socioeconomic status. Finally there is the chronosystem which is understood as the effects of the passage of time on development, not merely in a chronological sense, but in the sense of transitional events over the life course and historically. For example, one could expect the development of a child to differ in the wartime and post-war eras due to environmental differences brought about over time.

Social work practice has developed over time and the strategies used to protect vulnerable young people have adapted accordingly. Using the ecological model, we can examine how the approach to social work has changed. The original child rescue model, favoured by charitable and religious organisations in the 19th century, focused on removing a child from harm. In order to do so, the child was removed from their microsystems and placed into new systems which were presumed to be better. To a certain extent this has continued into more recent history with removal for adoption still being prevalent amongst young children in the care system (Barth and Berry 1988, DfE 2014). Displacing a child into an entirely new system was seen by many as a fresh start, although research has indicated that there may be some adverse effects in doing so (Lindholm and Touliatos, 1980). In particular, young people adopted from preschool age were more likely to suffer from conduct disorders and personality problems than their non-adopted peers. Furthermore, adopted children in a study in the USA were more likely to suffer poorer school adjustment between fourth and eighth grade than their peers (Ternay et al 1985).

Perhaps the 'natural' development for social work then was for the child to remain within the family unit for as long as possible. Where this was not possible, the child would be encouraged to retain contact with birth parents with an eventual aim towards reunification. The 1946 *Curtis Report* emphasised the importance of working with families in order to prevent abuse and neglect. This encouraged a shift in social work practice from a reactive force that displaced young people to a preventative force that acted to improve the nuclear family unit and strengthen the microsystems of the developing person. There are some advantages to this approach in that the systems are familiar and consistent for the young person. However, for many young people the risk of harm is too great for them to remain at home during the period of intervention with birth parents. It is for this reason that placements outside of the parental home still play a key role in social work practice.

Kinship care can perhaps be described as the middle ground for social work practitioners seeking to protect young people from harm. Placement with relatives enables practitioners to remove children from the potential dangers of the parental home whilst minimising the changes to their ecology. There are many elements and systems within a kinship placement that remain the same: continuity of educational establishment and peer groups, consistency in family relations and in many cases, due to the geographical proximity of placement to parental home, stability in neighbourhood (Ehrle and Geen 2002). As such, rather than displace the child, kinship care placements simply change the dynamics of the linkages between the microsystems. By doing so social work practitioners can work with the family to maintain the functional aspects of the child's ecology whilst reinforcing or adapting the dysfunctional aspects. Through this we would expect that kinship care, where appropriate and where there is suitable involvement from public services, would be superior to mainstream foster care.

A problem that is now recognised by the State is that of the intergenerational cycle of abuse (Buchanan 1998). This is a self-perpetuating cycle of care dependency in which parents who were in the care system themselves are more likely to re-enact abuse they suffered with their own children. For the State, kinship care may pose a partial solution to this problem. This is summed up in the proverb, 'Give a man a fish, and you feed him for a day. Teach a man to fish, and you feed him for a lifetime.' The child rescue model can be compared to the first part of this proverb in that it displaces a single child and fails to address the social issues that endangered that child in the first instance. This means that should another child be born, the same problems would present and the next child would also need to be rescued. Kinship care, however, offers an alternative solution. By investing resources into strengthening the ecology surrounding a vulnerable child, social work can create a self-sustaining and protective family unit. Essentially kinship care is about altering the dysfunctional elements to enable a family unit to be self-sufficient without the need for further intervention from the State. In terms of cost-effectiveness it is better to teach a family to care for itself in this way than to 'give them a fish'. By initiating these kinship support networks where appropriate, social work has the potential to break the self-perpetuating cycles of care dependency and in doing so reduce the number of children within the care system.

In some respects the prominence of formal kinship care placements within the care system is a fresh approach to out of home placements for young people. Despite a long anthropological and sociological basis for kinship networks and informal care relationships, the rise of kinship care in professional social work is a relatively modern phenomenon. However, it could be argued that the theoretical underpinnings of kinship care are anything but novel. Social work practice has undergone many changes with various trends in the approaches it has taken to child protection, but the sociological aim has always remained the same. At its core, the function of social work has always been to reinforce and support the family unit. From the child rescue model where the child's ecology was replaced in its entirety to the kinship model where the dysfunctional elements of the ecology are simply adjusted, the family has remained a central tenet of social work.

Conclusions

The theories outlined here indicate that social work has evolved as a reactive force responding to the breakdown of the nuclear family. Despite changes in the methods by which it acts, social work continues to serve its purpose in reinforcing the functions of the family unit. One could argue that this has historically been a latent function of social work; however, the growth of kinship care in practice may represent a growing recognition of this function within the field. Secondly, the theories highlighted in this chapter serve to generate debate surrounding our conceptual understanding of kinship and how it can be defined. We conclude that the biological explanations of kinship care are insufficient and that sociological theories of kinship organisation provide a more comprehensive account of kinship organisation. The altruism displayed by both kinship and kithship carers supports this theory and may have wider implications for policy at a local and national level. Finally, this chapter postulates that given suitable support and appropriate levels of intervention, kinship care has the potential to be beneficial for the State and the young people for whom they currently provide care. Where family-based placement options are available, kinship care provides a means to promote the self-sustaining family unit by strengthening resilience and in turn break the intergenerational cycle of care dependency.

References

Abrams, D.E. and Hogg, M.A. (1990). *Social identity theory: constructive and critical advances*. New York, Springer.

Barrett, L., Dunbar, R. and Lycett, J. (2002) *Human evolutionary psychology*. Princeton, Princeton University Press.

Barth, R.P. and Berry, M. (1988) *Adoption and disruption: rates, risks, and responses*. Piscataway, Transaction.
Biehal, N., Clayden, J., Stein, M. and Wade, J. (1995) *Moving on: young people and leaving care schemes*. London, HM Stationery Office.
Broad, B. (1998) *Young people leaving care: life after the Children Act 1989*. London, Jessica Kingsley.
Bronfenbrenner, U. (1977) Toward an experimental ecology of human development. *American Psychologist* 32, pp.513–32.
Bronfenbrenner, U. (1979) *The ecology of human development: experiments by design and nature*. Cambridge, Harvard University Press.
Bronfenbrenner, U. (1992) *Ecological systems theory*. London, Jessica Kingsley.
Bronfenbrenner, U. (1994) Ecological models of human development. *Readings on the Development of Children* 2, pp.37–43.
Buchanan, A. (1998) Intergenerational child maltreatment. In Danieli, Y. (ed.) *International handbook of multigenerational legacies of trauma*. New York, Springer, pp.535–52.
Buss, D.M. (2015) *The handbook of evolutionary psychology*. Chichester, John Wiley and Sons.
Cheney, D.L., and Seyfarth, R.M. (1992) *How monkeys see the world: inside the mind of another species*. Chicago, University of Chicago Press.
Daly, M. and Wilson, M. (1980) Discriminative parental solicitude: a biological perspective. *Journal of Marriage and the Family* 42, pp.277–88.
Dawkins, R. (1976) *The selfish gene*. Oxford, Oxford University Press.
DfE [Department for Education] (2014) *Statistical release: Children looked after in England (including adoption and care leavers) year ending 31 March 2014*. London, Department for Education.
DfE (2015) *Children looked after in England including adoption: 2014 to 2015*. London, Department for Education.
Eberhard, M.J.W. (1975) The evolution of social behavior by kin selection. *Quarterly Review of Biology* 50, pp.1–33.
Ehrle, J. and Geen, R. (2002) Kin and non-kin foster care: findings from a national survey. *Children and Youth Services Review* 24, pp.15–35.
Emerson, R. (1960) *From empire to nation*. Boston, Beacon Press.
Giddens, A., Duneier, M. and Appelbaum, R.P. (2003) *Introduction to sociology*. New York, W.W.Norton.
Hamilton, W.D. (1964a) The genetical evolution of social behaviour. I. *Journal of Theoretical Biology* 7, pp.1–16.
Hamilton, W.D. (1964b) The genetical evolution of social behaviour. II. *Journal of Theoretical Biology* 7, pp.17–52.
Haney, C., Banks, W. and Zimbardo, P. (1973) Interpersonal dynamics in a simulated prison. *International Journal of Criminology and Penology* 1, pp.69–97.
Home Office (2011) *Nature of bicycle theft. Crime statistics*. London, Home Office.
Huskinson, T., Kostadintcheva, K., Greevy, H., Salmon, C., Dobie, S., Medien, K., Gilby, N., Littlewood, M. and D'Souza, J. (2014) *Childcare and early years survey of parents 2012–2013*. London, Department for Education.
Lewin, R. (1974) Accidental Career. *New Scientist* 63(909), p.325.
Lindholm, B.W., and Touliatos, J. (1980) Psychological adjustment of adopted and nonadopted children. *Psychological Reports*, 46(1), pp.307–10.
Macionis, J., and Plummer, K. (2009) *Sociology: a global introduction*. Harlow, Pearson.
Merton, R.K. (1957) Manifest and latent functions. In Merton, R.K. (ed.) *Social theory and social structure*. Chicago, Free Press. pp.73–138.
Murdock, G.P. (1949) *Social structure*. Oxford, MacMillan.

Nandy, S., Selwyn, J., Farmer, E., and Vaisey, P. (2011) Spotlight on kinship care. Using Census microdata to examine the extent and nature of kinship care in the UK. Final report. Bristol: University of Bristol.

National Foster Care Association (1997) *Foster care in crisis: a call to professionalise the forgotten service*. London, National Foster Care Association.

Nowak, M.A., Tarnita, C.E. and Wilson, E.O. (2010) The evolution of eusociality. *Nature* 466, pp.1057–62.

OED [Oxford English Dictionary] (2016a) Altruism [definition 2]. Available at: www.oed.com/view/Entry/5857?redirectedFrom=altruism#eid

OED (2016b) Kinship [definition 1a]. Available at: www.oed.com/view/Entry/103585?redirectedFrom=kinshipand

OED (2016c) Kinship [definition 1b]. Available at: www.oed.com/view/Entry/103585?redirectedFrom=kinshipand

Parsons, T. (2013) *The social system*. London, Routledge.

Pope, W. (1975) Durkheim as a Functionalist. *Sociological Quarterly* 16, pp.361–79.

Rees, P. (2010) State dilemmas in the provision of alternative care for children: relative efficacy of public sector and independent sector foster placements. *International Journal of Public Administration* 33, pp.325–34.

Rees, P. (2013) The mental health, emotional literacy, cognitive ability, literacy attainment and 'resilience' of 'looked after children': a multidimensional, multiple-rater population based study. *British Journal of Clinical Psychology* 52, pp.183–98.

Scottish Government (2015) *Children's social work statistics Scotland, 2013–14*. Edinburgh, Scottish Government.

Selwyn, J. (2009) *EP19 Kinship Care from NICE guideline PH28*. Available at: www.nice.org.uk/guidance/ph28

Selwyn, J., Farmer, E., Meakings, S. and Vaisey, P. (2013) *The poor relations? Children and informal kinship carers speak out*. Bristol, School for Policy Studies.

Statistics for Wales. (2015) *Adoptions, outcomes and placements for children looked after by local authorities, year ending 31 March 2015*. Cardiff, Welsh Government.

Tajfel, H. and Billig, M. (1974) Familiarity and categorization in intergroup behavior. *Journal of Experimental Social Psychology* 10, pp.159–70.

Tajfel, H., Billig, M.G., Bundy, R.P. and Flament, C. (1971) Social categorization and intergroup behaviour. *European Journal of Social Psychology* 1, pp.149–78.

Ternay, M.R., Wilborn, B. and Day, H. (1985) Perceived child-parent relationships and child adjustment in families with both adopted and natural children. *Journal of Genetic Psychology* 146, pp.261–72.

Toyama, M. (2001) Adaptive advantages of matriphagy in the foliage spider, *Chiracanthium japonicum (Araneae: Clubionidae)*. *Journal of Ethology* 19, pp.69–74.

Waugh, I. (2014) *Children's social care statistics Northern Ireland 2013/14*. Belfast, Department of Health, Social Services and Public Safety.

Welsh Government. (2015) *Children in need at 31 March by looked after status, category of need and disability, including unborn children*. Cardiff, Welsh Government.

Williams, G. C. (1966) *Adaptation and natural selection: a critique of some current evolutionary thought*. Princeton, Princeton University Press.

Winokur, M., Crawford, G., Longobardi, R. and Valentine, D. (2008) Matched comparison of children in kinship care and foster care on child welfare outcomes. *Families in Society* 89, pp.338–46.

4 State of care

The ontologies of child welfare in British Columbia[1]

Sarah de Leeuw

Introduction

In January 2011, Mary Ellen Turpel-Lafond – a Cree lawyer and British Columbia's Representative for Children and Youth – presented *Fragile Lives, Fragmented Systems: Strengthening Supports for Vulnerable Infants* to the Legislative Assembly of British Columbia. The report reviewed the lives of 21 infants under the age of two who died between June 1, 2007 and May 1, 2009 while, in some capacity, living within the purview of the provincial Ministry of Children and Family Development (MCFD). MCFD is the governmental body responsible for child protection, family development, adoption, foster care, early childhood development and child care, child and youth mental health and justice services, and children and youth living with special needs in British Columbia. Of the 21 infants who died, 19 had lived with families served primarily by MCFD and two of the infants lived with families served solely by an MCFD delegated Aboriginal Agency.[2] Five infants were being cared for in a foster home or a respite care placement, the ministry-approved option provided to parents who voluntarily agree to take a break from parenting for a short time. Three of the 21 infants were, by law, 'in care'; the children had been involuntarily apprehended from their families and placed under governmental guardianship. "All of the infants in this review were born into families facing tremendous challenges" noted Turpel-Lafond (2011, p.3), but "an alarming number of the infants – 15 of the 21 – were Aboriginal".

This chapter considers the tragedies detailed in Turpel-Lafond's report and, more specifically, the ongoing challenges faced by Indigenous peoples in B.C. with regard to child welfare interventions into their lives. I propose the deaths of 15 Indigenous infants, all of whom were known to the governmental agency tasked specifically with family development and protecting children, can only be fully understood if considered within broader historical contexts and in reference to hegemonic and colonial systems. These systems are so fully naturalised as to be mostly invisible to especially settler-colonists (Morgensen 2011), but they have systematically marginalised Indigenous peoples across a wide variety of sociocultural and physical geographies in British Columbia and beyond (see

Clayton 1992, 2000, Kelm 1998, Sparke 1998, Harris 2002, 2004). I argue that a carefully produced, maintained, and historically contiguous set of logics premised on assumptions of neutrality, good intentions, and common sense vis-à-vis Indigenous children and families legitimises work done today by B.C. child protection services. As written about elsewhere in reference to ways that colonial powers are sustained (Waterstone and de Leeuw 2010), Gramsci's (1985) examination of common sense is useful for understanding colonial state hegemony: common sense has a powerful ability to transform and adapt to "scientific notions and philosophical opinions that have entered into common circulation" and to then rigidify "popular knowledge in a given time and place" (p.421). Indeed, as historian Ann Laura Stoler (2008) points out, those at the helm of colonial projects have always "understood, if unevenly, that their political rationalities were dependent on crafting sensory attunements and affective affiliations that made up common sense" (p.361). When common sense is called upon in the enactment of laws and policies, especially laws and policies which themselves are built upon colonial subjects' constructions about Indigenous peoples and places, the outcomes can be tragic. Common sense in systems of governance, as Derrida (2001) points out, is often maintained and aligned with languages about goodness, morality, and thoughtful considerateness (p.41). Destabilising these languages, and the ideas of common sense they convey, is an important step toward more socially-just systems of governance: "Language … the words themselves, are a first mediating institution. There is always a strategical or political calculation in the generous gesture of one who offers … amnesty, and it is necessary always to interrogate this calculation in our analyses" (ibid., p.42). Pulling apart the ostensible neutrality and the deeply affective nature of what appear to be commonsensically good intentions, languages, and practices provides insight into emancipatory possibilities elided by those with power, in this case those at the helm of governance systems that continue to violently bear down upon Indigenous peoples and the spaces of their families.

Fundamentally, then, I am concerned with disrupting hegemonic, common-sense, taken-for-granted ways that Indigenous peoples are conceptualised and consequently governed, principally by and within non-Indigenous systems of power, and thus – potentially – with ameliorating ongoing material inequities and injustices lived in Canada and beyond. To that end this chapter deconstructs applied practices and policies that marginalise Indigenous peoples, taking as its starting point that child welfare and child protection interventions into Indigenous families *are* ontological – despite existing at one register in discursive and epistemic realms, they are importantly and simultaneously lived and embodied, ultimately unfolding upon and with material consequence to particular kinds of bodies, families, homes, and communities. Child welfare, then, provides a productive frame through which to illustrate, with reference to colonialism's ongoing impact on Indigenous peoples, the inextricability of epistemic systems and ontological violences. This focus on child welfare systems is itself notable, particularly at a moment when geographers and other social scientists are examining the potential connections between

Indigeneity and ontology (broadly). A good deal of that work focuses on territory, land, and the environment, and associated questions of sovereignty, development, economics, or resources (see for instance Braun 2002, Castree 2004, Johnson and Murton 2007, Louis 2007, Wainwright 2008, Blaser 2009). This is not out of synch with traditional divisions between the epistemological 'world as known, as thought' versus the ontological 'world as is, as being' that have driven Human Geography's broader, although not fixed, positioning of ontological investigations within, principally, nature/culture divides or relationships between human society and nature, place, and space (Woodward and Jones 2009, pp.511–3). Human Geographers' conceptualisations of ontologies are shifting, notably resulting in growing critique of hard and fast binaries between subject/object and knowing/being, with an attendant interest about concepts of 'becoming' (Woodward and Jones 2009). In research about Indigeneity and Indigenous geographies, however, the literature remains dominated by considerations about Indigenous cultures and identities flowing from physical environment or slippages between human and non-human objects/environments within Indigenous ways of knowing and being (see for instance Bunkse 2007, Watson and Huntington 2008). Although I suspect these conceptualisations should and could be usefully critiqued for continuing resilient histories of framing Indigenous peoples and Indigeneity as inextricably linked to land, physical territory, the environment or, more problematically, a wilderness counterpart (see Cameron et al 2009), I am more concerned in this chapter with expanding what has to date simply been a lack of geographic inquiry about the ontologies of, and attendant anti-colonial and political possibilities in, micro-scale, diminutive, vernacular, and domestic places lived by Indigenous children and their families – many of which are led by Indigenous women. By tracing historic common-sense attitudes about Indigenous people within epistemic and discursive realms, and by demonstrating that child protection and welfare policies are not as neutral, given, or indeed commonsensically good and well-intentioned either as they might first appear or as governmental officials suggest them to be, I am interested in the ways that epistemology, discourse, and ontology work *together* to ensure an ongoing taken-for-granted colonial supremacy wherein Indigenous families and communities continue to be marginalised. Child welfare, then, can be productively theorised through frameworks offered by feminist geographers (Rose 1993, Massey 2005) and others who recognise ontologies as relational, as co-constitutively produced by people and spatialities, and as always informed by (especially unequal) relationships of power: as Ann Laura Stoler (2008) astutely summarises about colonial relations, especially in domestic and diminutive domains, "epistemic considerations are neither transcendent nor abstract. They are of the colonial world and squarely in it" (p.351). There is merit in framing child welfare, the apprehensions of Indigenous children, and the geographies of Indigenous families and communities, as ontological concerns. There is merit too in aligning such analysis with broader ontological efforts in geography and beyond to interrogate colonialism and disrupt hegemonic structures that continue to undermine social justice for Indigenous peoples, in British Columbia and beyond.

Children at risk need protecting: it has always made good common sense

Implementations of child protection and child welfare services, which involve government-sanctioned interventions into families, are invasive and disruptive acts (Fournier and Crey 1997, Walmsley 2005, Foster and Wharf 2007). Child apprehensions, one outcome of assessments that conclude a child is at risk, involve social workers, and sometimes police officers or security guards or hospital and school officials, removing children from their families, often without parental or child consent. First-person testimonials by parents about their experiences of having their children removed rely on words like 'stealing', 'baby snatching', 'kidnapping', and 'life altering' (Jones 2010, see also Walmsley 2005, Pivot Legal Society and Law Foundation of British Columbia 2008). The work surrounding an apprehension consists of time-consuming and arduous processes involving social workers, courts, lawyers, family mediation programs, and parental supervision by state authorities, all of which can be devastating to a family. As one mother explains, "[s]ince the day of my daughter's birth, I have been fighting to keep the MCFD from removing her: doing as required by the MCFD [means] proving the accusations against me wrong [and] putting my own life (education) on hold. It sure isn't easy" (Jones 2010, unpaginated). Like other state actions that suspend considerations granted to the majority of a population or to a majority of individuals (e.g. imprisoning a person), intervening into families for child protection purposes requires systems of logic and law that provide clarity and justification about why implementing the invasive, disruptive, and sometimes even violent practices is, fundamentally, justifiable or, at least better than not implementing them (Blomley 2004). Child protection services, and the social workers who implement the services, are charged with balancing the conflicting pressures of, on the one hand, society's wish to protect children from abuse and, on the other hand, broad social agreement that the family unit is a bastion of liberty which should be intervened into only with utmost caution (Munroe 1996, Regehr et al 2002).

Given that British Columbia joined the confederation of Canada in 1871, legalised state intervention into families has a relatively long history in the province. In March 1901, the first child welfare act was passed in the BC Legislative Assembly, solidifying in law the benevolent efforts of the Council of Women of Vancouver to ensure that children were not left in circumstances and places where they faced abuse, neglect, or abandonment (Foster and Wharf 2007). Child welfare practices and policies were premised on the idea that the spaces of some families were so antithetical to children's well-being that it was incumbent upon the state to remove them and (re)place them into safer, healthier, spaces. Logics about child welfare rested upon the assumption that there were objective quantifiable means of assessing the goodness of some families' domestic spaces versus the corruptness or unwholesomeness of others. Assessments were often based on the hygiene of a home, morality of the parents, future potential afforded to the children by living in the home, and behaviours of parents and relatives towards each other and the children (see for example Canadian Council for Child Welfare 1927).

The impulse to assist and better the lives of those living on the margins of society, and in Canada this has often included Indigenous peoples, is remarkably immune to critique at the time of its unfolding (de Leeuw et al 2010). With ostensibly all good intentions, non-Indigenous government, non-governmental, and ecumenical agencies have a long history of constructing themselves as protecting and benevolent forces aiming to 'improve' and 'help' Indigenous peoples 'for their own good' (Haig-Brown and Nock 2006). Much of this work was underwritten by logics of common sense:

> The epistemic tasks that colonial institutions faced were contingent on forms of knowledge that bureaucratic machines … uneasily embraced: problematic predictions about the political consequences of people's affective and moral states … [depended on] … a shared common sense about what was likely, allowing prediction and directing the political projects it served. (Stoler 2008, p.353)

Common sense was (and is) a crucial colonial strategy with remarkable resiliency, offering colonisers a means to distance themselves from more obviously coercive or violent ideas, actions, and policies. Protecting, helping, and keeping Indigenous children and families safe is a project with a long history in Canada and British Columbia (de Leeuw 2007, de Leeuw et al 2010). As early as 1845, the government of the then-not-yet Canada voiced interest in the welfare of those considered most at-risk for deleterious and a-moral behaviors, Aboriginal peoples: "[We are] interested in the welfare of this race … [of] mass[ing] valuable information upon their present state, and suggestions for improving it … to form a judgment upon any scheme proposed for their future management" (Rawson et al 1845, p.3).

These policies and laws formed the foundation of what is now referred to by some as Canada's 'national crime' (Milloy 1999) and recognised by many others as perhaps one of the nation's most egregious interventions into Indigenous peoples' families and social structures. Residential schooling, although it shifted depending on the times and places in which it was delivered, was an effort, as outlined in federal government policy documents and program documents, to "produce Indians … adjusted to modern life [and] capable of meeting the exacting demands of modern society with all its complexities" (see Canada, Department of Indian Affairs Branch 1947). From its onset, residential schooling was framed as a means of 'saving' Indigenous children, through education – which had high moral standing in Canada throughout the 19th and 20th centuries – from the inherent lack of education, civility, safety, and progress understood to mark the spaces of Indigenous families and communities. Residential schooling was presented as a benevolent commonsensically good endeavour concerned with child welfare and protection (Miller 1997, Milloy 1999, de Leeuw 2007, 2009).

Common sense benevolence, then, is not a response to a politically neutral assessment of 'need' among Indigenous peoples. Rather, it is a longstanding contributor to a cycle of identifying deficiencies and needs that can then be 'fixed,' making control more palatable to those involved. Historically, there has been a constant (re)production of Indigeneity – and places it is associated with – as abnormal or as normally aberrant (Cameron et al 2009). Spaces and bodies of Indigenous peoples are rarified and codified in policy and law as unique in ways that require the good intentioned interventions of non-Indigenous powers. For Indigenous peoples, this cycle involves an endless deferral – often by governments – of the time when they can be deemed 'ready' to 'manage' themselves (de Leeuw et al 2010). For over 150 years, discourses comprised of laws, policies, and popular rhetoric about the need to protect and care for those unable to do so themselves have had disastrous results for Indigenous people, including severe inequities in health, standards of living, and overall individual and community well-being (Government of British Columbia 2002, Adelson 2005, Assembly of First Nations 2006).

Today's *Child Welfare Act*, and those who now enforce it, has undergone various iterations through the early decades of the 20th century. Early child protection workers were mostly charity workers or members of religious benevolent societies. Child welfare policies and practices saw their most significant transformations between 1930 and 1960. During that time, social work became an institutionally recognised university degree in the province, the depression left increasing numbers of families economically marginalised, and Indian residential schools, which through the 1950s were the principle places and means for the state to 'care' for Aboriginal children (de Leeuw 2007, 2009), began to shut down (Foster and Wharf 2007). The 1960s through the 1980s saw what has become known as 'the sixties scoop' (Johnson 1983): in 1955, as Indian residential schools began to close in the province, 29 Aboriginal children entered the child welfare system; by 1964 the number had risen to 1,446 (Foster and Wharf 2007). Numbers of Aboriginal children entering the provincial child welfare system have grown steadily (and increased as a percentage of the total population) since the 1960s (Hughes 2006). In 2008, over 9,200 children were in care in British Columbia (Pivot Legal Society and Law Foundation of British Columbia 2008). The percentages of Aboriginal children now in care have been called epidemic and calculated as worse than the numbers of children in residential schooling (Walmsley 2005, Assembly of First Nations 2006, Hughes 2006).

The Ministry of Children and Family Development (MCFD), who in 2012 was tasked with child welfare and protection in the province, operates within the legal framework of the *Child, Family, and Community Service Act* (*CFCS Act*), passed in the Legislative Assembly of B.C. in 1996 and amended through to include 2011. Since 1996, the CFCS Act, MCFD personnel, and the Ministers appointed to lead MCFD, have changed so many times that, according to more than one commission tasked with evaluating the ministry and the success or failures of its mandates, the lack of any kind of consistency is unto itself compromising the ability of the state to deal safely with child welfare (Hughes 2006, Foster and Wharf 2007, Pivot Legal Society and Law Foundation of British Columbia 2008). Between 2004 and 2011, with a growing number of children dying either while in the care of MCFD or after having not received governmental care despite calls for such care, and in the midst of growing and sustained criticism, the provincial government has undertaken significant efforts to explain child welfare and protection services to British Columbians writ large. The results are a plethora of reports, websites, and 'family-friendly' materials designed to explain MCFD logics and practices. The advent and proliferation of these publications results in a cohesive messaging about good, common-sense, ways of assessing if a child requires the care of the state.

According to these publications, the basic premise behind MCFD's work is to provide "programs and services to ensure that healthy children and responsible families are living in safe, caring and inclusive communities" (Government of British Columbia MCFD 2007a, unpaginated). In February 2011, when Mary Polack was the MCFD Minister, she observed that "[o]ur government is committed to ensuring that all children and youth in British Columbia have the strongest start possible. Safety is always of the utmost importance and is always our number one priority.... Keeping children and youth safe is the responsibility of all British Columbians" (Polack 2011, unpaginated).[3] After ministerial changeover later in 2011, the fifth since 2006, the new Minister of MCFD Mary McNeil observed that

> [t]his is an exciting time of change for British Columbia ... [with] a new agenda that is putting B.C. families first ... [u]nder the 'families first' vision ... [an] action plan [that] puts a strong focus on early intervention and a needs-based approach to supporting and protecting vulnerable children and youth – through a strong, integrated system. (Government of British Columbia MCFD 2011b, p.3)

Posters for child abuse prevention hotlines read "It's Everyone's Responsibility to Keep Kids Safe. You Can Help". The *Child Family Community Service Act* requires by law that "anyone who has reason to believe that a child may be abused, neglected, or is for any other reason in need of protection, must report it to the Director or a delegated social worker" (Government of British Columbia MCFD 2011a). People who suspect a child is being abused or neglected are encouraged to call a provincial hotline to report their suspicions.

On the surface, it is difficult to negate these publications' messaging: who would want or advocate for unhealthy children living in unsafe, uncaring and non-inclusive spaces? Who would argue against children having a strong and safe start to life? These publications, then, produce a set of common-sense answers or referents to questions about child welfare. Concepts like 'healthy children' and 'responsible families' however, are not always easily assessable, definable, quantifiable, or immune to the sociogeographic and historical contexts in which they are circulated. Indeed, given that for many Aboriginal families in B.C. the times are less exciting than suggested by Minister McNeil, it seems productive to think critically about the ways that common sense is (re)produced through languages that frame families as needing to be healthy and responsible while, simultaneously, Indigenous peoples continue to experience disproportional sufferings that stem from colonial incursions into lives and lands.

Responding to Child Welfare Concerns: Your Role in Knowing When and What to Report (Government of British Columbia MCFD 2007a), a booklet that abounds in schools, doctor's offices, and social service agencies, outlines indicators that a child may be being physically, sexually, and/or emotionally abused. Indicators that might arouse suspicions of child maltreatment are disturbing, clear, and obvious; they range from lesions, visible bruising, and hand or object marks on any/all parts of a child's body through to severe antisocial behaviours and suicidal ideations. The factors that would elicit suspicion of child abuse seem reasonably commonsensical: an injury on a child "in the shape of a stove burner or a belt buckle" (ibid, p.1) is cause for concern, worthy of reporting to officials and professionals trained to reach well-reasoned decisions about such situations. The booklet also clarifies that "you do not need proof. Just report what you know.... If you do not want to give your name or phone number, that is okay" (ibid, p.3). The indicators leave little doubt that it makes good and common sense to report any inclination that a child may be being abused. The booklet's description of neglect, which it is also mandatory to report, is less clear:

> Neglect happens when a parent or guardian ignores or overlooks a child's basic needs – to the point where the child is, or could be, harmed. Neglect includes failing to provide a child with food, shelter, basic health care, supervision, nurturing or protection from risks. (ibid, p.5)

Family relatives or other concerned citizens are offered indicators of neglect, including "[c]lothing that does not protect the child from the weather ... [and/or] poor personal hygiene" (ibid, p.6). The language about neglect is more nuanced than the language about abuse, but informational materials disseminated by MCFD provide on the whole a unified, clear, and apparently commonsensical suggestion that child welfare and child protection is good, that everyone in a population is vested in child safety and the maintenance of healthy and responsible families, that people can report suspicions in anonymous ways, and that partaking in the surveillance of children's well-being (or lack thereof) means partaking in a morally helpful and legally mandated practice.

Once a child welfare concern is reported, assessing the concern and deciding on next steps is immediate:

> if there are concerns about a child's safety, a child welfare worker has a legal duty to respond under the *Child Family and Community Service Act*. The child welfare worker has a number of options after making an assessment: take no further action, refer the family to support services, ... evaluate the family's strengths and needs and ... develop a plan to provide support services, [or] conduct a child protection investigation. (Government of British Columbia MCFD 2011a)

Even if no further action is deemed necessary, a case file is opened about the reported family and the particulars of the complaint are documented and filed. Should it be concluded that action of any kind is indeed required, the actions are taken by social workers who are guided by the *Child and Family Development Service Standards* (CFDSS) manual, referred to colloquially as 'the practice standards' manual (as per personal communication with MCFD in 2011). The manual's purpose is

> to promote consistent high-quality service for vulnerable children, youth and families while enabling flexibility in the way services are provided to meet the unique needs, capacities and resources of communities.... The standards are not based on a particular service delivery model. Rather, they are designed to be flexible enough to work within a variety of service delivery approaches. (Government of British Columbia MCFD 2009)

Language is carefully considered and explained in the manual, indicating strong linkages between linguistic and discursive formations of family issues, child welfare concerns, and the practices and procedures deployed to address them:

> With the exception of the administrative procedures, there is minimal use of procedural language in the standards. The language focuses on the **intended outcomes of the service, rather than process or the steps to reach the outcomes**.... The approach taken in the standards affords greater flexibility in practice and allows the standards to fit a variety of service delivery models currently existing in communities and in the future. (ibid, p.9)

The most up-to-date service standards governing assessment of whether or not a child is at risk in B.C., and thus guiding subsequent interventions, emphasise 'broadness' and 'descriptiveness' while simultaneously lacking specificity on applied steps and practices or concrete means that might allow a social worker to ascertain if abuse or neglect exists in a family. The CFDSS manual refers social workers with questions to the *Handbook for Action on Child Abuse and Neglect: For Service Providers* (Government of British Columbia MCFD 2007b) that, beyond very obvious indicators such as bodily harm, starvation, and sexual exploitation of children, provides little clarity on the nebulous situations bound to present themselves when assessing the need for child protection. Despite the handbook noting "[c]hildren do not always tell us about their abuse or neglect, and sometimes the indicators are not obvious", the processes to assess if a child is in need of protection are entirely absent (ibid, p.39). Nowhere in the CFDSS manual or associated handbooks is any mention made to the spaces or places where concerns might arise or where assessments might occur. No discussions are offered about ways that a social worker might understand different geographies beyond their own frames of reference or expertise (including for instance, the geographies of Indigenous families whose lives and realties are inextricably intertwined with experiences of colonialism). Indeed, there are no references to how Indigenous communities with disproportionate levels of poverty and sociocultural trauma, which are strongly evidenced as the outcome of colonisation by the very systems making governance decisions about Indigenous peoples and places, might be thoughtfully evaluated on their own merits or terms as requiring (or not requiring) state intervention. Embodied or place-specific languages to guide assessments and/or subsequent actions and interventions are also nowhere to be found. Instead, social workers are encouraged "not to dwell on procedures" but, rather, to "reach outcomes".

While reinforcing that social workers must always be mindful of 'community', the publications offer no concrete or site-specific indicators about the place where assessments and interventions will unfold. The languages and discussions are placeless and disembodied, offering no guidance to those implementing law and policy regarding child welfare and protection about how to account for the active nature of sociospatial contexts when rendering their judgements and upholding laws. Guided either by the CFCS Act or CFDSS manual, a social worker is instead instructed that (i) when screening reports and establishing approaches to service delivery, the 'intent' is to acknowledge that "a careful assessment involves the family and community, taking into account their strengths, needs, culture, and beliefs" and (ii) all MCFD policies "are based on an assessment of the context and needs of the child and family" (Government of British Columbia MCFD 2004). No languages specify "contexts" as produced by a colonial history or a colonial present. There are no languages or guidelines offering insight into the ways a social worker, a child welfare worker, or even a concerned citizen who fears child neglect may be unfolding, can account for colonialism's ongoing dispossession of Indigenous communities – which results in poverty and a host of other marginalised realities – while simultaneously taking into account the strengths, needs, culture resiliencies, and beliefs of Indigenous families and communities. The MCFD consultant responsible for training child protection and child welfare social workers in the largest health region of B.C. (the Northern Health Authority) observed there is "no ticky-box tool" or any concrete or definitive set of criteria used by social workers to assess if a child requires protection; instead, a social worker "makes their assessments based on their experience and training" (pers comm with MCFD in 2011). Social workers are encouraged to consider the need for stability and continuity of life-long relationships in a child's life, a concept that is vague and open-ended enough that little choice would be left other than to bring into play one's best, commonsensical, interpretation and judgement. Both the CFSC Act and the Service Standards manual make mention of assessments and decisions being 'community specific' which, according to the MCFD trainer in northern B.C., means that social workers are encouraged to "remember where they are" when forming judgements: it is important, for instance, that a social worker "not have the same" or universal expectations for "housing and general living standards" in different communities because a rural or remote town might be "really different" than an affluent "urban place" (MCFD, pers comm 2011).

Details or concrete descriptions that could inform community or place-specific assessments, however, are not offered.

The CFCS Act, the CFDSS manual and other MCFD publications about child protection and welfare are full of wording about Aboriginal peoples and, insofar as the publications make regular mention of "Aboriginal culture", are full of opaque references to Indigenous identity, heritage, community, and Indigeneity. Nowhere, however, are there any details about the deleterious outcomes of colonial dispossession of Indigenous communities, about which much evidence abounds (see for instance Harris 2002, 2004). No linkages are drawn between historical and contemporary colonial interventions into families and territories and the corresponding sociocultural breakdowns that are, globally, understood as being linked (Adelson 2005). This is despite the fact that, in the last decade, all the materials have undergone significant revisions so that wording conveys governmental concern with the 'best interests' of Indigenous children, families, and communities. Indeed, the CFDSS manual now has independent and stand-alone standards that apply only to Aboriginal families and communities; one of the seven 'key themes' of the CFDSS manual is "keeping Aboriginal children connected with their families and communities, and strengthening collaborative working relationships with Aboriginal communities" (Government of British Columbia MCFD 2004). One of the seven 'guiding principles' of the law upon which all child welfare and protection practices are based in B.C. (the CFCS Act) is that "the cultural identity of aboriginal children should be preserved" (Government of British Columbia MCFD 2011a). One of the five 'service delivery principles' of the law is that "aboriginal people should be involved in the planning and delivery of services to aboriginal families and their children"; of the two subsections in the act devoted to the 'best interests of the child', one section is simply that "[i]f the child is an aboriginal child, the importance of preserving the child's cultural identity must be considered in determining the child's best interests" (Government of British Columbia MCFD 2011a). Despite the language, however, the numbers of Indigenous families into whom the state intervenes for child protection and welfare purposes continues to grow. And the numbers of infants and children who die while in the purview of MCFD, as Turpel-Lafond (2011) writes, is alarming.

The reasons Indigenous families are being reported, investigated, and intervened into are not, on the main, based on physical or sexual violence and abuse about which there is such clear language in publications about child welfare. In four cycles (1998, 2001, 2003 and 2008) of the *Canadian Incidence Study on Reported Child Abuse and Neglect* commissioned by the Public Health Agency of Canada (PHAC 2010), *neglect* is the leading reason that children – particularly Aboriginal children – are brought to the attention of child welfare and protection services (see Blackstock et al 2004, Bennett et al 2005). In 1998, reports of neglect comprised 40 per cent of investigations into child maltreatment, 57 per cent of which were found to be unfounded (PHAC 2010). These numbers were virtually unchanged in 2001 and shifted only slightly in 2003 (Trocmé et al 2005). According to the 2008 Incidence Study, of the over 235,000 child maltreatment investigations that occurred in Canada, 30 per cent of all the investigations were found to be based on unfounded concerns. This does not undermine the severity and seriousness of the 70 per cent of substantiated cases, of which 'neglect' and 'exposure to intimate partner violence' (a new category in 2008) comprised an equal majority (at 68 per cent) of the incidences of child maltreatment at 34 per cent respectively (PHAC 2010). These numbers, however, do suggest that full clarity about what constitutes child maltreatment does not exist in those who report it and that suspicion of child neglect remains a primary reason for people to report suspicions of child maltreatment and, consequently, the primary reason that families are investigated and/or intervened into. Given the deeply subjective nature of ascertaining what constitutes neglect, common sense likely factors heavily into cases of neglect being reported and, subsequently, into the ways the cases are evaluated and understood.

The ontologies of child protection in British Columbia

Child protection and child welfare practices in British Columbia are, today, operating within a paradox. On the one hand, the child protection law states both that the CFCS Act "must be interpreted and administered so that the safety and well-being of children are the paramount considerations, that children are entitled to be protected from abuse, neglect and harm or threat of harm" and that "a family is the preferred environment for the care and upbringing of children and the responsibility for the protection of children rests primarily with the parents" (Government of British Columbia MCFD 2011a). Also, according to the CFCS Act, every effort must be made to ensure that "kinship ties and a child's attachment to the extended family ... be preserved if possible and ... the cultural identity of aboriginal children should be preserved" (Government of British Columbia MCFD 2011a). On the other hand, two things are still happening in British Columbia – with little sign of abating. First, Aboriginal families are, for reasons of suspected child neglect, being intervened into by MCFD at disproportionally higher rates than any other families in the province (Assembly of First Nations 2006, Hughes 2006, Pivot Legal Society and Law Foundation of British Columbia 2008, Turpel-Lafond 2011). Secondly, disproportionally high rates of Aboriginal children, once under the purview of MCFD because of concerns about neglectful domestic spaces, are meeting with tragedy, including death (Assembly of First Nations 2006, Turpel-Lafond 2011). Practices, polices, and laws full of common-sense language focused on the safety, health, and well-being of Aboriginal children and families are resulting in just the opposite. As of 2008, Aboriginal children made up 51% of children in care in BC; the province's 2008/2009 Auditor General Report noted that "[s]tudies indicate that in British Columbia, an Aboriginal child is about 6 times more likely to be taken into care than a non-Aboriginal child. Of all B.C. children who are in care, 51 per cent are Aboriginal — yet Aboriginal people represent only about 8 per cent of B.C.'s population" (Office of the Auditor General, British Columbia 2008). The 2010 MCFD service plan reported that 53 per cent of 8,677 children in the ministry's care were Aboriginal (Government of British Columbia MCFD 2010) and the 2008

Canadian Incident Studies Report on child protection notes a rate of 49.69/1000 of Aboriginal families reported for investigation compared to only 11.85/1000 reports on non-Aboriginal families (Trocmé et al 2010). These are particularly pressing issues because the health and well-being of Aboriginal communities writ large has repeatedly being linked to the health and well-being of their children and Aboriginal peoples as a whole when compared with non-Aboriginal peoples in Canada are, into the 21st century, still living with some of the lowest standards of health and well-being in the country (Health Canada 2008). For Indigenous peoples there is an undeniable link between child protection interventions and an overall compromised standard of life – indeed, the more families are disrupted by state interventions, the more fragile they become (Walmsley 2005).

What appear to be good, commonsensical, state policies about child welfare and child protection practices require interrogating as ontological issues. Addressing and deconstructing child protection practices, logics, laws, and policies that are being called forth, enacted, and enforced by people relying on common sense (that is steeped in discursive terrains that have always (re)produced Aboriginal peoples as uniquely and aberrantly racialised and requiring salvation, mostly from themselves by a well-intentioned state) must occur in tandem with understanding and addressing child protection as an embodied, material, and ontological issue that always takes place *in place*. Child protection imperatives always exist within a lived world wherein drastic disparities are experienced by Indigenous children and families, disparities that themselves flow from a suite of prior colonial interventions and assumptions. To recognise and privilege an ontological orientation to child protection requires troubling the ascendance of common-sense languages about good and moral domestic spaces, which are burgeoning and dominating the conversations about child welfare, particularly because the spaces of far too many Indigenous families are sites of people living racialised lives marked by poverty. The ascendance of common-sense languages about what children need must constantly be tempered with concurrent discussions about addressing and critically understanding the realities of Indigenous peoples living in geographies marred by contemporary colonialism. Common sense does work and is active in the service of colonial power; undoing contemporary colonial power requires destabilising commonsensical ideas about what constitutes good child welfare practices and in whose best interests those practices are being undertaken while simultaneously emplacing the languages within a material framework. Thus, while on the surface it makes common sense that all children and youth in British Columbia have the "strongest start possible" and that making this so provides for an "exciting time" that supports and protects vulnerable children and youth in British Columbia, such discursive frameworks may be precisely what results in certain domestic spaces becoming vulnerable for public scrutiny and state intervention, rendering them ontologically further at risk of the violent interruptions that result from child protection services.

The spaces of some Indigenous families reflect anything but exciting times. Turpel-Lafond (2011) notes that all 21 infants who died before the age of two years old, while involved with MCFD, between June 1, 2007 and May 1, 2009, were born into families facing "tremendous" challenges. Serious poverty and inadequate housing were common to all cases. Nine of the mothers were either single or very young parents. Five were first-time mothers. Twenty of the 21 families had intergenerational trauma in the parent's family of origin and 15 of the 21 families had a documented history of abuse or neglect in the mother's family of origin. Fourteen of the 21 families had domestic violence issues in the immediate family. Eight of the families had documented domestic violence in previous generations. Sixteen of the 21 families had substance abuse in their immediate family, and 15 of the 21 had substance abuse in previous generations. Twelve of the 21 families had documented mental health issues such as depression, suicidal behaviour, and anxiety. Common sense, once (re)produced in epistemic and discursive spaces, takes place and is enacted some*where*, thus always being informed and constituted within the sociospatial and temporal contexts where it unfolds. In B.C., anyone being investigated for child abuse or neglect undergoes a "fact-finding process [wherein] a child welfare worker sees and interviews the child in the household, directly seeing the child's living situation" (Government of British Columbia MCFD 2007b, p.5).

Where a person resides, including their home and their broader community, and the state of being from the perspective of those sites from the perspective of a professional, are active forces in the making of decisions about indicators of child maltreatment, including that nebulous concept of neglect. Indigenous peoples across Canada live in what some call "third world conditions of health" and what others refer to as the "embodiment of inequality" (Adelson 2005). While Canada is ranked among the best places to live in the world, if the United Nations Human Development Index was applied to Indigenous peoples living on-reserve, the country would rank between 68th and 80th in the world (Webster 2006). Aboriginal peoples in Canada have higher rates of disease and die younger and at higher rates than their non-Aboriginal counterparts (Allard et al 2004). Their life expectancy, although increasing, still falls 'well below' that of non-Indigenous peoples; rates of tuberculosis are 8 to 10 times higher in First Nations communities in comparison to non-Indigenous communities and other infectious diseases such as hepatitis, chlamydia, and HIV/AIDS are all either more prevalent or increasing at higher rates in First Nations communities as compared with non-Indigenous populations (Adelson 2005). In 2006, nearly one-fourth of all Indigenous people in Canada lived in homes in need of major repairs and, when compared with their non-Indigenous counterparts, Indigenous peoples were almost four times more likely to live in a crowded dwelling (Trocmé et al 2008). In northern interior B.C., First Nations peoples directly correlate the Indian reserve communities in which they live, spaces they were relocated to because of settler colonial policies that did not establish treaties when deterritorialising Indigenous peoples in the province, with being racialised in specific ways and being consequently discriminated against: as one First Nations woman noted,

> I'd like people to know that just because we're Indians, we're not a lower life form. Which is really what you get.... It's a fact of life, we all know it. (de Leeuw et al 2012, unpaginated)

The Aboriginal families who attracted the gaze and attention of MCFD between June 1, 2007 and May 1, 2009 were absolutely multiply-marginalised, opening them up to scrutiny through a variety of normative sociogeographic lenses which can find families or communities wanting and unfit or, conversely, that consider potential markers of risk as simply 'par for the course' in the rarefied and already 'not normal' geographies lived in and experienced by marginalised, poor, or suffering families. If both the general public, who is called upon to report suspicions about child maltreatment, and professionals, who are called upon to judge if child maltreatment is occurring, scrutinise realities of these families in part through common-sense orientations about what constitutes good and moral domestic spaces, about which people are being actively schooled through multitude publications circulated by the MCFD, it makes sense that interventions might appear appropriate in the case of impoverished, marginalised, or racialised Indigenous families.

The very idea of family is a slippery, and quickly evolving, socially constructed concept. As the deluge of publications about at-risk children demonstrates, however, there are repeating refrains about what constitutes 'good' families in and for British Columbia – these refrains are rarely ontological, skirting the material injustices and inequities lived by Indigenous peoples. Instead, publications entrench and reify common-sense and commonplace conceptualisations about where and how children should be raised. Children should be in safe places. Children should be raised in homes and communities where they are well cared for, nurtured, dressed for the seasons, and provided with appropriate health care and educational opportunities. They are sheltered and part of 'healthy communities'. Such testaments, however, continuously sidestep the ontological, remaining materially opaque and without reference to a concrete lived world. Domestic spaces and families not adhering to these commonsensically good standards become, by default, wrong or bad domestic spaces and families. In their aberrance, they are transformed into rarified geographies in need of help and care.

Conclusions

Removal of Indigenous children from homes, families, and communities has occurred for a long time. Rarely have the logics that underpinned these removals not turned on benevolence and a 'common sense' of the day: it once made good, moral, sense (at least to those with the power to enforce it) to save souls, to civilise 'savages' who lived as families in places perceived as filthy and sinful, and to impart markers of modernity like literacy and education upon people for whom assimilation was understood as inevitable. It is difficult to argue against the imperative of keeping children – and all people, for that matter – safe and free from abuse. And it is true that children are particularly vulnerable members of any society. Despite the vast, government-produced, rhetoric about making B.C. safe for all children and families, about a happy and healthy future for all British Columbians, Aboriginal families are still facing disruptive, violent, state government intervention at unparalleled rates. Despite the disproportionally high rates of apprehensions and interventions, Indigenous children are still dying once under the purview of the MCFD. Two things may be precipitating this. First, because of deeply ingrained ideas about Indigenous people being aberrant and requiring intervention, Indigenous children are seen – and thus reported – as requiring more care, more often, than are non-Indigenous children. More Indigenous families being consequently investigated, whatever the outcome of the investigations, (re)produces expectations about the elevated nature of Indigenous families' needs in terms of care by the state. Secondly, with an ascendance of rhetoric about what constitutes good and morally sound families, more attention may be focused on domestic places that seem not to adhere to common-sense ideas about what constitutes safe, strong, exciting, and healthy domestic spaces. Indigenous children and families are, by default and by the logics and rhetoric that collude to reproduced them as such, perceived as always abnormal and unsafe, thus always requiring intervention or, because of racialised expectations about what is 'normal' for them in their 'abnormality', simply being ignored. Constructed and contextualised as such, a child living in genuinely dangerous circumstances may be written-off within the confines of what might be expected from 'normal' and 'good' families, thus not worth deep and sensitive inquiry.

Curtailing this cycle rests upon carefully and purposefully deconstructing languages and rhetoric about child and family well-being, particularly how common sense is called upon during the interpretation and consequent implementation of policies and laws governing child welfare. Understanding that child welfare is an ontological and a geographic issue is also needed. Relationships between families, children, home or domestic places, and child welfare policies are neither abstract nor too diminutive to hold remarkable potential in efforts to destabilise settler colonial power. Despite dwelling and circulating, somewhat at least, in the epistemic realm, child welfare practices are material and embodied: they must be evaluated not as abstract, but as material, lived, and in-place violences that, ultimately, impact lives in the real world. This requires understanding – and making ontological – the fickle nature of concepts like caring, nurturing, safe homes and families or, conversely, ideas about some families innately not being as safe or as caring as others and thus 'normal' in that state. It requires holding these realisations in constant reference to understandings about the outcomes of colonisation – as opposed to producing more and more opaque rhetoric about common-sense ways to conceptualise and evaluate intimate and diminutive spaces lived by children and families. Policies and laws need to be explicit about the materiality of child welfare assessments and interventions. This means careful wording and concrete descriptors for those who enact government policies, replete with tools about how to disrupt common sense and subjective responses to cases of child welfare and child protection. Like all people, those in positions of power to monitor and assess concepts like neglect and child welfare (thereby finding some families failing and wanting or leaving children in places where they are at risk) bring with them a set of spatially and temporally – a set of ontologically – informed expectations.

These expectations, particularly when they unfold in places that are new or different or involve people who are racialised in particular ways, might not in the end make much 'common' sense at all. This, I think, has consistently proven and continues to be the case for Aboriginal people in B.C. The results are tragic.

Acknowledgements

This chapter was previously published as de Leeuw, S. (2014) State of care: the ontologies of child welfare in British Columbia. *Cultural Geographies* 21, pp.59–78. We thank Sage Publications for permission to republish under license number 3854450276476 (April 2016). Minor edits have been made to align the manuscript with the house style of this edited collection. Research for this paper, and related projects focused on colonial projects in British Columbia, was funded by the Social Science and Humanities Research Council of Canada (SSHRC), number 410-2010-1413. I am very grateful for the thoughtful feedback of three anonymous reviewers. The efforts of Emilie Cameron and Caroline Desbiens are similarly much appreciated.

Notes

1 This chapter was previously published as de Leeuw, S. (2014) State of care: the ontologies of child welfare in British Columbia. *Cultural Geographies* 21, pp.59–78. See Acknowledgment for details.
2 A delegated Aboriginal Agency is a "designated representative" according to the *Child, Family Community Service Act*: i.e. a "treaty first nation, an Indian band or aboriginal community" (including the independent Nisga'a Lisims Government) that has been designated, in accordance with regulations, to implement all aspects of child protection and child welfare law (see Government of British Columbia, Ministry of Children and Family Development [hereafter MCFD], 2011a, chapter 46).
3 In March 2011, Polack was reappointed as Minister of Aboriginal Relations and Reconciliation. She was replaced by Mary McNeil.

References

Adelson, N. (2005) The embodiment of inequalities: health disparities in Aboriginal Canada. *Canadian Journal of Public Health* 96, pp.45–61.

Allard, Y., Wilkins, R. and Berthelot, J. (2004) *Premature mortality in health regions with high Aboriginal populations*. Ottawa, Statistics Canada.

Assembly of First Nations (2006) *Leadership action plans on child welfare*. Ottawa, Assembly of First Nations.

Bennett, M., Blackstock, C. and de la Ronde, R. (2005) *A literature review and annotated bibliography on aspects of Aboriginal child welfare in Canada*. Ottawa, First Nations Research Site of the Centre of Excellence for Child Welfare and the First Nations Child and Family Caring Society of Canada.

Blackstock, C., Trocmé, N. and Bennett, M. (2004) Child maltreatment investigations among Aboriginal and non-Aboriginal families in Canada. *Violence Against Women* 10, pp.901–16.

Blaser, M. (2009) The threat of the Yrmo: the political ontology of a sustainable hunting program. *American Anthropologist* 111, pp.10–20

Blomley, N. (2004) Law, property, and the geography of violence: the frontier, the survey, and the grid. *Annals of the American Association of Geographers* 91, pp.121–41.
Braun, B. (2002) Colonialism's afterlife: vision and visuality on the northwest coast. *Cultural Geographies* 9, pp.202–47.
Bunkse, E. (2007) Feeling is believing, or landscape as a way of being in the world. *Geografiska Annaler: Series B* 89, pp.219–31.
Cameron, E., de Leeuw, S. and Greenwood, M. (2009) Indigeneity. In Kitchin, R. and Thrift, N. (eds.) *International encyclopedia of Human Geography, volume 5*. Oxford, Elsevier, pp.352–7
Canada, Department of Indian Affairs Branch (1947) *The Indian School Bulletin* I-IX.
Canadian Council on Child Welfare (1927) Report of the British Columbia child welfare survey. Ottawa, Canadian Council on Child Welfare.
Castree, N. (2004) Economy and culture are dead! Long live economy and culture! *Progress in Human Geography* 28, pp.204–26.
Clayton, D. (1992) Geographies of the lower Skeena. *BC Studies* 94, pp.29–58.
Clayton, D. (2000) *Islands of truth: the imperial fashioning of Vancouver Island*. Vancouver, University of British Columbia Press.
de Leeuw, S. (2007) Intimate colonialisms: the material and experienced places of British Columbia's residential schools. *Canadian Geographer* 51, pp.339–59.
de Leeuw, S. (2009) 'If anything is to be done with the Indian, we must catch him very young': colonial constructions of Aboriginal children and the geographies of Indian residential schooling in British Columbia, Canada. *Children's Geographies* 7, pp.123–40.
de Leeuw, S., Greenwood, M. and Cameron, E. (2010) Deviant constructions: how governments preserve colonial narratives of addictions and poor mental health to intervene into the lives of indigenous children and families in Canada. *International Journal of Mental Health and Addictions* 8, pp.282–95.
de Leeuw, S., Maurice, S., Holyk, T., Greenwood, M. and Adam, W. (2012) With reserves: geographies of colonialism and First Nations health in northern interior British Columbia. *Annals of the American Association of Geographers* 102, pp.904–11.
de Leeuw, S. (2014) State of care: the ontologies of child welfare in British Columbia. *Cultural Geographies* 21, pp.59–78.
Derrida, J. (2001) *On cosmopolitanism and forgiveness*. London, Routledge.
Foster, L. and Wharf, B. (2007) *People, politics, and child welfare in British Columbia*. Vancouver, University of British Columbia Press.
Fournier, S. and Crey, E. (1997) *Stolen from our embrace: the abduction of First Nations children and the restoration of Aboriginal communities*. Vancouver, Douglas and MacIntryre.
Government of British Columbia (2002) *Report on the health of British Columbians. Provincial Health Officer's annual report 2001: the health and well-being of Aboriginal people in British Columbia*. Victoria, Ministry of Health and Planning.
Government of British Columbia MCFD [Ministry of Children and Family Development] (2004) *British Columbia's family development response*. Victoria, MCFD.
Government of British Columbia MCFD (2007a) *Responding to child welfare concerns: your role in knowing when and what to report*. Victoria, MCFD.
Government of British Columbia MCFD (2007b) *The handbook for action on child abuse and neglect: for service providers*. Victoria, MCFD.
Government of British Columbia MCFD (2009) *Child, family, and community services act*. Victoria, MCFD.
Government of British Columbia MCFD (2010) *Children in the care of the Ministry of Children and Family Development in the north: the third annual monitoring progress report*. Victoria, MCFD.
Government of British Columbia MCFD (2011a) *Child, family and community service act [RSBC 1996]: Chapter 46*. Victoria, MCFD.

Government of British Columbia MCFD (2011b) *Ministry of Children and Family Development, revised 2011/12–2013/14 service plan*. Victoria, MCFD.

Gramsci, A. (1985) *Selections from cultural writings*. Cambridge, Harvard University Press.

Haig-Brown, C. and Nock, D. (eds.) (2006) *With good intentions: Euro-Canadian and Aboriginal relations in colonial Canada*. Vancouver, University of British Columbia Press.

Harris, C. (2002) *Making native space: colonialism, resistance, and reserves in British Columbia*. Vancouver, University of British Columbia Press.

Harris, C. (2004) How did colonialism dispossess? Comments from an edge of empire. *Annals of the Association of American Geographers* 94, pp.165–82.

Health Canada (2008). First Nations and Inuit Health Branch. Fact Sheet, http://www.hc-sc.gc.ca/ahc-asc/branch-dirgen/fnihb-dgspni/fact-fiche-eng.php.

Hughes, T. (2006) *BC children and youth review: an independent review of BC's child protection system*. Victoria, Minister of Children and Family Development.

Johnson, P. (1983) *Native children and the child welfare system*. Toronto, Lorimer.

Johnson, J. and Murton, B. (2007) Replacing native science: indigenous voices in contemporary constructions of nature. *Geographical Research* 45, pp.121–29.

Jones, J, (2010) *BC authorities snatch three-day-old indigenous baby*. Available at http://vancouver.mediacoop.ca/story/bc-authorities-snatch-three-day-old-indigenous-baby/4303

Kelm, M. (1998) *Colonizing bodies: Aboriginal health and healing in British Columbia, 1900–1950*. Vancouver, University of British Columbia Press.

Louis, R. (2007) 'Can you hear us now?' Voices from the margin, using indigenous methodologies in geographic research. *Geographical Research* 45, pp.130–9.

Massey, D. (2005) *For space*. London, Sage.

Miller, J. (1997) *Shingwauk's vision: a history of native residential schools*. Toronto, University of Toronto Press.

Milloy, J. (1999) *A national crime: the Canadian government and the residential school system, 1879–1986*. Winnipeg, University of Manitoba Press.

Morgensen, S. (2011) *Spaces between us: queer settler colonialism and indigenous decolonization*. Minneapolis, University of Minnesota Press.

Munroe, E. (1996) Avoidable and unavoidable mistakes in child protection work. *British Journal of Social Work* 26, pp.793–808.

Office of the Auditor General, British Columbia (2008) *Management of Aboriginal child protection services*. Victoria, Ministry of Children and Family Development.

PHAC [Public Health Agency of Canada] (2010) *Canadian incidence study of reported child abuse and neglect-2008: major findings*. Ottawa, PHAC.

Pivot Legal Society and Law Foundation of British Columbia (2008) *Broken promises: parents speak about B.C.'s child welfare system*. Vancouver, Law Foundation of British Columbia.

Polack, M. (2011) *Message from the Minster*. Victoria, Government of British Columbia.

Rawson, W., Davidson, J. and Hepburn, W. (1845) *Report on the affairs of the Indians in Canada: submitted by Charles Bagot to the Legislative Assembly*. Ottawa, Province of Canada at Kingston.

Regehr, C., Chau, S., Leslie, B. and Howe, P. (2002) Inquiries into deaths of children in care: the impact on child welfare workers and their organisations. *Children and Youth Services Review* 24, pp.885–902.

Rose, G. (1993) *Feminism and geography: the limits of geographical knowledge*. Minneapolis: University of Minnesota Press.

Sparke, M. (1998) A map that roared and an original atlas: Canada, cartography, and the narration of nation. *Annals of the Association of American Geographers* 88, pp.463–95.

Stoler, A. (2008) Epistemic politics: ontologies of colonial common sense. *Philosophical Forum* 39, pp.349–61.

Trocmé, N., Fallon, B., MacLaurin, B., Daciuk, J., Felstiner, C., Black, T., Tonmyr, L., Blackstock, C. and Barter, K. (2005) *Canadian incidence study of reported child abuse and neglect – 2003: major findings*. Ottawa, Minister of Public Works and Government Services.

Trocmé, N., Fallon, B., MacLaurin, B., Sinha, V., Black, T., Fast, E., Felstiner, C., Hélie, S., Turcotte, D., Weightman, P., Douglas, J. and Holroyd, J. (2008) Canadian Incidence Study of Reported Child Abuse and Neglect. Ottawa, Ontario: Public Health Agency of Canada.

Trocmé, N., Fallon, B., MacLaurin, B., Sinha, V., Black, T., Fast, E., Felstiner, C., Hélie, S., Turcotte, D., Weightman, P., Douglas, J. and Holroyd, J. (2010) *Canadian incidence of reported child abuse and neglect – 2008: major findings*. Ottawa, Minister of Public Works and Government Services.

Turpel-Lafond, M. (2011) Fragile lives, fragmented systems: strengthening supports of vulnerable infants. Victoria, Representative for Children and Youth.

Wainwright, J. (2008) *Decolonizing development: colonial power and the Maya*. Oxford, Blackwell.

Walmsley, C. (2005) *Protecting Aboriginal children*. Vancouver, University of British Columbia Press.

Waterstone, M. and de Leeuw, S. (2010) A sorry state: apology excepted. *Human Geography* 3, pp.1–28.

Watson, A. and Huntington, O. (2008) 'They're here – I can feel them': the epistemic spaces of indigenous and western knowledges. *Social and Cultural Geography* 9, pp.257–81.

Webster, P. (2006) Canadian Aboriginal people's health and the Kelowna deal. *The Lancet* 368, pp.275–6.

Woodward, K. and Jones, J. (2009) Ontology. In Gregory, D., Johnston, R., Pratt, G., Watts, M. and Whatmore, S. (eds.), *The dictionary of human geography*. Oxford, Blackwell, pp.511–3.77

5 Caring citizens

Emotional engagement and social action in educational settings in New Zealand

Bronwyn E. Wood and Rowena M. Taylor

Introduction

The use of educational spaces to regulate children and young people in the ways of modern citizenry is not new. Schools can be thought of as an aggregation of the values, aspirations and ideals held by social and political agents in society, who employ a range of strategies to attempt to mould young citizens in certain ways (Staeheli 2011). Citizenship education programmes are key to such endeavours as they provide a strategic site for the creation of dispositions and characteristics which will engender responsible, 'good' citizens in the future. In recent years, a growing number of countries have developed and extended their focus on citizenship education (Nelson and Kerr 2006, Brooks and Holford 2009, Schulz et al 2010, Kennelly and Llewellyn 2011). Whilst the reasons for this heightened focus on citizenship education are complex, Nelson and Kerr (2006) suggest that new social, economic and environmental challenges and declining rates of youth civic participation have contributed to this. Increasingly, these citizenship education programmes require more 'active' expressions of citizenship where young people not only need to know *about* civic processes, but also to *participate* as active citizens (Ross 2012). This has led to a many new policy initiatives that encourage young people to participate in acts of citizenship – such as through service learning, community participation, social action and volunteering – in many countries including the UK, the US, Canada, Australia and New Zealand (Farahmandpour 2011, Davies et al 2012, Jerome 2012, Birdwell et al 2013, Wood and Black, 2014).

However, while more active forms of citizenship education have been adopted across many countries, very little attention has been given to how young people actually *feel* about undertaking citizenship acts as part of their compulsory curriculum. Emotions such as empathy, compassion, concern or anger are likely to underpin the desire or inspiration to perform citizenship acts (Jasper 1998), yet, within the context of citizenship education, little mention is made of these emotions and how integral they are to citizenship action – especially within formal curricula policy. Instead, citizenship education curricula often appeal to a distant, abstract sense of loyalty to a nation founded upon notions of 'belonging to community' (Wood 2013) or 'responsibilities' and 'duties' as citizens of a democracy (Kennelly and Llewellyn 2011). Moreover, no guidance is given about what teachers or students are to do in the absence of a sense of care or commitment about the required acts of citizenship or if young people do not feel like performing these acts. Furthermore, taking social action also relies on a sense of agency which leads citizens to act on behalf of others (Osler and Starkey 2005). This is also closely linked to emotions such as confidence and qualities such as persistence. Yet, there is strong evidence that young people frequently feel disempowered in society as a result of marginalising practices and discriminating processes in society (Kennelly and Dillabough 2008) – an aspect citizenship education policy rarely takes into account. This then raises a number of questions in this context: What happens when young people are 'required' to care about an issue in order to take citizenship action as part of curricula or assessment? What if they don't care? Or if they don't feel empowered to act? Does this make them 'bad' citizens? And, how do teachers navigate these requirements?

This chapter examines the interplay of care and young people's social actions within an active citizenship education programme in New Zealand schools. While others have previously proposed that 'caring' is required to make the world a better place (Massey 2004), or that caring itself can be a political act (Bartos 2012), the intersection between emotions of caring and acts of citizenship in schooling has not been well explored. Furthermore, when research has connected emotional engagement with civic action, it has tended to come from psychologically-derived research that fails to account for the nuances of place, temporality and human relationships. If we understand that emotions are relational, and that they arise and flow between people (Bondi 2005), we need to examine how caring occurs within the intergenerational relationships and socio-spatial interactions represented within active citizenship education contexts. In doing so, this research also contributes to wider discussions about the role and nature of emotions in societal interactions and in social activism (Brown and Pickerill 2009), the significance of empathy in citizenship and social justice endeavours, and the implications of this on the type of citizen evoked within curriculum and educational policy (Pykett et al 2010).

The context for this research is New Zealand, where, since 2011, students who study Social Studies in the final three years of secondary schooling (Years 11–13, ages 15–18) are assessed for undertaking 'personal social action'. This national assessment initiative forms part of New Zealand's credentialing system, contributing to the National Certificate of Educational Achievement (NCEA) (NZQA 2013). Social Studies has traditionally been the primary vehicle for citizenship education in New Zealand (Archer and Openshaw 1992) and now includes this social action component at senior levels for both curriculum and assessment purposes. This chapter draws from a two-year research project (2015–2016) examining the experiences of teachers and students from five New Zealand secondary schools who were undertaking these social action achievement standards. While curriculum and assessment documents do not include reference to the affective dimensions of taking personal social action, it became evident very early on in our project that for students to take action, they needed a sense of emotional engagement or connection with the selected social issue. In short, they needed to 'care'. Hence, in this chapter, we examine the affective dimension of social action from both the teachers' and students' perspectives, to explore the significance and nature of 'care' in an active citizenship process.

The chapter begins by providing a conceptual overview for understanding emotions drawing upon Zemblyas' (2007) tri-fold conceptualisation of theoretical approaches to emotion in educational contexts. The aim of this analysis through Zembylas is to consider how theoretical approaches to emotion shape our understandings of emotion in active citizenship. Drawing on empirical data from our research project, we examine how teachers fostered the conditions for 'caring' and 'care-full' classrooms, through their own affective responses to social issues and pedagogical strategies, and their students' responses. Two groups of students are examined – those who 'cared' and those who didn't care (as much). We discuss the implications of these affective positions, and examine how emotions such as caring interact with acts of citizenship. We conclude by considering the implications of these findings, arguing that meaningful social action needs to be centred on an authentic cognitive and emotional connection to social issues, and that teachers need to be aware of their role in nurturing this affective engagement if they are to encourage the development of active and empowered citizens.

Caring classrooms: how do we understand care in educational contexts?

Care is inseparable from educational contexts. Children and young people in educational institutions have always been 'cared *for*'. Much less attention, however, has historically been paid to exploring how young people themselves are carers, or how classrooms are 'sites of caring'. One reason why emotions in classrooms are often an 'absent-presence' (Boler 1999, p.xv) is the long-standing priority given in education to cognitive pedagogies which are reasoned, logical and as 'uncontaminated' by emotion as possible (Kenway and Youdell 2011). Thus emotions are often relegated to a backseat in favour of truth, rationality and reason, or are simply unseen (Sheppard et al 2015).

A recent growing body of work addresses this imbalance. In education, Boler's (1999) seminal work in *Feeling Power: Emotion and Education*, and Zembylas' exploration of the politics of emotion in classrooms (Zembylas 2005, Zembylas et al 2011, 2014) have helped to advance thinking in this area. Alongside this, a growing literature that specifically addresses the emotional geographies of education has highlighted the inseparable nature of emotions from the social, cultural, economic and political landscapes of education (Kenway and Youdell 2011, Nairn and Higgins 2011, Bartos 2012, Wood 2013, Ahmed 2014, Black 2015, Gagen 2015). These studies of emotions have helped to interrogate the "embodied, material and particularized experience of our daily lives" (Boler 1999, p.7) in educational contexts and draw our attention to the deep connections between emotions, people and places (Bondi et al 2005). Importantly, these recent studies have shifted the focus on emotions in education beyond neurological, psychobiological and psychological foci to a broader, social constructivist understanding which views emotions as historically and socially constructed as well as deeply shaped by our relations with others.

Schools are particularly important sites for citizen formation as they have long played a key role in establishing rules, norms and behaviours in order to create citizens capable of functioning in particular ways (Marston and Mitchell 2004, Staeheli 2011, McLeod 2012, Wood 2015b). A growing focus on emotional education has seen the launch of emotional literacy programmes in many countries with the aim of teaching students the skills of respectful social interaction, safe and healthy behaviours, enhanced self-esteem and ethical responsibility to their peers and community members (Boler 1999, Greenberg et al 2003, Gagen 2015). These programmes exemplify initiatives which aim to cultivate democracy through the promotion of certain emotions, of which empathy is the most popular (Boler 1999, Thrift 2004). As Boler (1999, p.156) states, "across the political and disciplinary spectrum, conservatives and liberals alike advocate for variations of empathy as a solution to society's 'ills'". Empathy emerges as a vital emotion for students to develop in order to appreciate how society's ills impact on other people's lives.

Studying empathy or care, similar to other emotions, however, is a complex endeavour (Sheppard et al 2015). Emotions are difficult to see, interpret and define. Moreover, a successful emotional performance can be artificially produced in order to serve a purpose or fit a specific context (Lindquist 2004, Hochschild 2012). Emotions can stem from 'transitory responses' to an external event (such as anger, indignation, fear, joy) or from deeper 'underlying affects' (including loyalty to family, friends or places, or fear of others) which contribute to deep affective attachments (Jasper 1998). Both forms of emotion shape politics, identities and citizenship responses as these are forged in and through social relations that rest upon deep feelings of commitment to people and places (Massey 2004). As emotions have moved more toward the centre of public life in recent years (Bondi 2005), in what Thrift (2004) calls an ever increasing intensity of affect, it is more important than ever to consider how they shape the political action of both young people and adults. In particular, we need greater understandings of the "sociality of emotion" (Ahmed 2014, p.8) in which emotions are understood to be located not just inside or outside people, but circulate between people, within social contexts, serving as a form of cultural politics or world making. A key challenge that this chapter addresses therefore is to examine the significance of affect in youth citizenship action in the context of citizenship education programmes in schools and to consider how such socio-emotional and spatial contexts shape citizen identities and actions.

The complex nature of emotional research meant that we needed to clarify the theoretical approach underpinning our study. We found Zembylas' (2007) three categories of approaches to the study of emotions a useful analytical tool to both understand and interpret how emotions form and circulate within socio-spatial contexts.

The first and arguably most dominant traditional approach to the study of emotions in education are *psychodynamic* approaches which are derived from psychological understandings of emotion. Emotion in this sense is largely understood as an internal feeling that relates to the reactions of individual subjects in response to external stimuli (Zembylas 2007). The second approach views emotion as a *sociocultural* experience. Also known as a social constructivist approach, understandings of emotions in this view are dependent on the context in which they are expressed and this will differ between cultures and contexts (Zemblyas 2007).

Zembylas (2007) defines a third approach as an *interactionist* position which views emotion as neither simply psychological nor social, but suggests emotions are "crucial to the processes in which the psychological and the social are produced" (p.63). This approach takes a closer look at sociocultural contexts and considers how emotions are formed at the intersection of human interaction. As Wetherell states, emotion and affect are part of relational practice: "… affect is distributed. It is an in-between, relational phenomenon" (2015, p.158). She explains the need to integrate the psychological with the sociological: "… we cannot split between semi-conscious, automaton-like, reactive body and the reflective, discursive, interpreting, meaning-making, communicating social actor. Equally we cannot isolate and reify affect" (p.160). Using this tri-fold framework, we explore how the emotions of care and empathy are constituted in the context of Social Studies classrooms and how these contexts mediate expressions and experiences of social action.

Methodology

Five schools were purposively selected for the study (Berg 2007) based on the experience and expertise of five teachers in these schools who were identified as leaders in this area and committed to undertaking social action with their students. A convergent parallel mixed method approach was taken in order to provide a comprehensive analysis of the research problem (Creswell 2014). This included a survey of Senior Social Studies teachers (n=145) alongside an in-depth case study approach with teachers and students in the five selected schools. Qualitative data collection included individual and focus group interviews with the five teachers involved in the project, cycles of action research as well as classroom observations of students planning and undertaking social action. The social actions take on average five weeks with students generally forming groups to select and inquire into a social justice issue, and to plan and undertake action(s). The focus of the formal assessment is on students' conceptual understandings and the clarity of their individual reflections upon the strengths and weaknesses of their approach and the impact of their actions, not directly on their emotional engagement. Self-selected students from all five schools took part in focus group discussions after they had completed their social action requirements.

This chapter draws primarily on data from the student focus group interviews and to a lesser extent on teacher interviews. The focus groups consisted of five to nine students each (*n*=52), providing an opportunity for students to reflect on and evaluate the experience collectively with others from the class (Berg 2007). These data were analysed to identify key themes and patterns that emerged within and between the five case study schools (Creswell 2014).

Creating the conditions for 'caring' classrooms

As noted earlier, the role that emotions play in the classroom is rarely the central focus of scholarship. Initially, the significance of emotion in taking social action went unseen (Sheppard et al 2015) in setting up our research. Our classroom observations and discussions with teachers alerted us to this crucial link between emotional engagement, agency and action in the social action process. Teachers were very clear at the outset that one of their first goals in the programme was to select social issues that would engender an emotional connection for their students if the social action was to be authentic and meaningful. Empathy was the most desired emotion, in keeping with Boler's (1999) claim that this is the most popular emotion to cultivate in democratic education. As Mel stated: "[I need to get my] students to identify or empathise with injustice, therefore become engaged in the context." Ann described how she spent hours on YouTube before introducing a social issue to the class "trying to find the clip that I know will hit them in the heart". Olivia stated:

> You just don't go "aw, I'm gonna have a social action just now". It's like a whole thing, the emotion that there's an injustice or something that needs working on. What will I do about it? What will we do about it? How will we do it? What should we do?

These quotes illustrate the integral connection between caring about an issue and emotional engagement and social action that teachers made in their teaching. Teachers saw their attempts to create a 'caring' atmosphere in their classrooms as part of wider goals they held of "creating an equitable society" (Olivia) and concerns about social justice (Mel). They saw the key avenue to this was by encouraging their students to "step into other people's shoes" (Janine) and by "helping young people feel connected, understand themselves, and [learn to] empathise with others" (Mel). These values underpinned their commitment to setting the conditions in their classrooms for meaningful social action with their students, as well as their own identities as social activists.

In order to support this emotional connection, teachers used a number of strategies to introduce social issues that they felt would be compelling. For example, they chose topics with high levels of affective content, showed video clips and documentaries that evoked empathy, and highlighted websites profiling children and young people who had made a difference. One teacher, for instance, showed a documentary about poverty in New Zealand and the poor quality of housing for some families. Another examined the plight of refugees fleeing the Middle East to

Europe by profiling personal blogs of refugees. Another showed videos of young people who were working against poverty to "'ra-ra' them up a bit" (Rachel). A key goal of these strategies was to create empathy for the people involved and the social issue focused on, resulting in a sense of agency to be able to undertake appropriate social action.

Through these practices, teachers demonstrated what Hochschild (2012) refers to as 'emotional labour', or the management of visible signs of emotion to create an atmosphere or an attribute which has exchange value. Their actions were committed to nurturing a 'caring' classroom. One teacher, Rachel, illustrated this by describing how she modelled emotions she wanted from the students – in this case indignation and outrage:

> I often I get angry about [social issues]. And I'll see something and just go "grrr!" in the front of the class and I'm "this so annoys me!" [My aim is to let them] see *your* emotion and then it's alright for them.

Lindquist (2004) suggested that staging empathy is one way to meet the goal of deeper forms of affective engagement from students. In Rachel's case, her modelling of emotions was designed to evoke students' greater emotional connection with the social issue studied. These strategies alert us to the socially constituted nature of emotions such as empathy, and how teachers fostered undeclared emotional 'rules' which served to delineate a zone within which certain emotions were permitted while others were not (Zembylas 2005). In the next section, we examine how students responded to these strategies to consider how such conditions were perceived and understood by students.

"We did it because it was quite personal": students who cared

The majority of students we observed and interviewed across the five schools demonstrated high levels of emotional engagement with their social action focus. For many, a connection or attachment to a social issue was felt emotionally and many voiced a visceral response to their chosen social issue, especially younger students whom we interviewed (aged 15 and 16). For example, in the following group discussion, three 14-year-old female students describe how a sense of empathy sparked a desire to take action:

> When we saw the living conditions that people, that some people are living in *damp homes* … it is just really horrible and I just couldn't imagine living in that sort of place … it was really horrible. I just wanted to help them.
>
> I think the more we learned about [child poverty in New Zealand] the more we felt how much we take things for granted. We learned that what we have is quite a lot you know, yet, like some kids don't really have anything
>
> Yeah … I really wanted to help. I really want to contribute like to bettering our society and stuff.

Others talked about how 'amazed' they were that an issue hadn't 'been noticed' by the public or how 'shocked' or 'moved' they were by events which seemed unfair or unjust. These responses were both 'transitory emotions' as well as indicative of deeper 'underlying effects' (Jasper 1998) which motivated their interest and empathetic responses. Young people described the citizenship that these assessments conveyed in terms of being caring, empathetic, active, involved, and/or 'walking the walk' in order 'to make a difference'. Importantly, underpinning this was a sense of compassion, as Josh (16 years) described:

> [They want us to be] empathetic. Not just know knowing how other people in less fortunate situations feel, but actually *feeling* it. Like moving you to wanna do something. Like not just being sympathetic and understanding, "oh yeah, they're suffering", but actually going "that's terrible, we need to do something".

Another described this type of citizen as a "good Samaritan" (Luke, 16 years), drawing on a Biblical reference to explain someone who possessed the agency to go out of his/her way to help those in need.

When it came to selecting their social issue, students were mostly motivated by issues which had a direct relevance to their lives and were situated in the 'micro-politics' of their local communities (Harris and Wyn 2009). Older students (17-18 years) in one focus group who had completed a social action assessment the previous year reflected that they "formed a better connection to [the local social issue] rather than [the national-level issue] we did this year – there's not that same level of connection". The significance of young people's experiences of places, communities and local issues in shaping their citizenship actions and dispositions is well established (Marsh et al 2007, Weller 2007, Harris and Wyn 2009, Harris et al 2010, Wood 2015a). Local experiences and knowledge also significantly influence young people's interest in social issues and the extent to which they are informed about current issues (Holden et al 2008). For example, having refugee students in the school was a reason some students chose to focus on this issue as "we thought it was quite important" (Angela, 16 years). Others similarly chose issues such as child poverty as "many students in our school go without food and lunch and … [we knew about that]" (Jess and Nadine, 14 years).

There were also examples of deeply personal choices of social issues to undertake social action on. Two boys (Jackson and Dylan, 14 years) in one school had a burning issue which affected their families. Jackson's mother, who was Dylan's aunt (both boys happened to be in the same class), had died in the public hospital a year previously. It had become apparent that over-crowding and understaffing had contributed considerably in the death of Jackson's mother and this had been confirmed by a story run by a local television station. The boys decided to write to the local hospital to ask for a formal apology and to address improvements in their systems of patient care. Their close connection to this event was obvious:

JACKSON: "My social issue was the way patients were being treated in the hospital. We did it because it was quite personal. [It was about what] happened in the past … "

DYLAN: "The reason why I did it was because it was personal to me as well and I believe that the company that we're arguing with [the hospital] didn't give it too much action [sic]."

Their social action was underpinned by a sense of injustice as well as personal tragedy as Dylan outlines. These examples illustrate how young people's "embeddedness in their local world shapes their thinking about politics and their political and social action" (Harris and Wyn 2009, p.329). These examples illustrate the integral link between emotion and a continuum of responses from altruism and good neighbourliness through to a sense of agency and more transformative 'activist' sensibilities seeking social justice. Not all students, however, experienced such empathetic responses. In the following section we examine students who didn't 'care' – or at least not as much as the ones described above.

"We weren't very passionate …": students who didn't care (as much)

In every school a small group of students undertaking the social action assessments did not engage deeply with the social issues and the resulting social action process. In some cases, these students were in their last year at school, and had basically 'checked out' already, and were not attending classes or completing school work. None of these largely absent or disengaged students self-selected into our focus groups. Others, however, participated through the social action process, but expressed some reluctance or frustration about this process – often as a result of not being very 'passionate' about the selected social issue. For example, in one class, Kimberly, a 17-year-old, described how her group of five students "wasn't very passionate about anything in the local community" which their teacher had insisted had to be the scale of focus. She described how she would have preferred a more political national-scale issue that was significant to more people. Two younger students described how they didn't really 'care' about their issue, but "it grew on us I guess and we cared more about it" (Jess and Nadine, 14 years). Andrea (16 years), reported that she did care about the issue but felt "outside her comfort zone" when it came to taking social action and while she did "feel good" once the action was completed, she would not take the subject the following year.

The degree of freedom within which social action was undertaken had a significant impact on young people's responses. Where there was less choice and freedom, students felt restricted and frustrated. For example, in one class where the teacher had pre-selected the social issue, Ben (17 years) felt it was "kind of imposed on us" and as a result said, "I don't feel as emotionally charged about it". Another group of students (aged 16-17 years) considered that the fact the social action was assessed "kind of dumbed it down a bit, because [...] it's not so much like you are personally involved, it's sort of like you've got ulterior motives to doing it". Ben's comment highlights the potential for some forms of coercion to potentially exist in citizenship classrooms, all in the name of creating 'good' citizens (see Wood 2015b for further on this). Teachers were aware of the potential to be criticised for 'social engineering' and most put in place some strategies to make sure students had some freedom throughout the process by selecting their own social issues and/or forms of social action. One teacher, Mel, went as far as saying, "I will never impose my choice of social issue on them – I am willing for them to fail the assessment before I'd do that".

One frustration voiced by students in all focus groups was how they had to find ways to manage the disengagement of other students from their group. Generally students completed their action in groups of between two and six, enabling them to spread the workload. Anjali (16 years) summed up the frustration of many when she described it as "kind of like carrying a dead weight":

> It was quite annoying ... some people were really unreliable. We had this plan that we had six actions that were to be done it was like video, emailing, posters, presenting at assembly, four of the actions didn't get done.

This necessitated students themselves 'managing' their peers in order to complete the social action and the assessment, including "giving them a job that they had to do. And we were constantly checking up on them and making sure they were doing that job" (Tania, 16 years). Josh described how he employed a strategy of asking his group members for their opinions "because there was no way they were going to offer their opinion without being prompted. So, we constantly asked them ... kind of forced them [laughs] to do stuff." In order to cope with students who were absent or disengaged, Hayley (16 years) said:

> We gave them a job that wasn't so important, that if they didn't do it the whole thing wouldn't go down, but they felt like they were still part of the group. And then if they didn't do it, we could still have done it before the end of the thing. So it worked out well and they were still part of it.

These discussions illustrate how students themselves had become part of the governing processes of the classroom, undertaking a fair amount of 'emotional labour' in their strategies to manage their peers' lack of engagement. This also illustrates the impact of a small group of students' emotional disengagement rippling out beyond the private feelings of individuals to shape the responses of their peers. These peers responded with a series of tactics which matched the hidden 'rules' of these classrooms contexts in a way that connects the psychological to the sociological with almost "seamless feedbacks" (Wetherell 2015, p.152).

Discussion and concluding thoughts

Care is a complex concept and there are multiple ways we can interpret what is happening in classrooms where care is required as part of a formal assessment regime. Our analysis has highlighted that care was an essential part of citizenship action. As Bartos noted, "care is an action, a practice, a goal and a strategy" (2012, p.160) and can be seen in how children and young people work to maintain, preserve and repair their worlds. Our research confirms the strategic importance of empathy and caring about a social issue if young people are to experience meaningful engagement in their acts of citizenship. Young people's sense of agency and engagement was significantly enhanced when faced with issues with which they had some level of personal experience, interest and understanding, a point made by Freire (1973) many years earlier. Connecting with young people's own experiences increased the sense of care and commitment and also reduced the potential for emotional engineering. In contrast, the students who were disengaged or lacked passion about an issue did not have that sense of emotional connection – either through their own or the teachers' choices – and their disengagement also impacted negatively on others.

Our research also illustrated that geographies of emotion were constituted within the fabric and social settings of classrooms and communities. As Simonson (2005, p.8), (drawing on Lefebvre's (1991) understanding of the temporality of everyday life) reflects, "emotions and affections, private life and its symbols cannot submit to cumulative and linear processes". Instead, the process of taking social action was dynamic, unpredictable and subject to multiple contingencies relating to people and situations. An analysis through Zembylas' (2007) theoretical framework confirms the dynamic and multiple layers of emotion in the citizenship classroom, and helps to explain the *psychological*, *sociocultural* and *interactionist* nature of emotions in this context. Each of these theoretical analyses of emotion added a layer of depth to our understandings of how emotion interacted with the active citizenship process.

The study found that an initial encounter with social issues triggered a *psychodynamic* or psychological reaction, such as a sense of care, empathy, or indignation and/or outrage. These responses were visceral, affective and often closely linked to agency and the desire to take action. Osler and Starkey (2005) suggest that first and foremost, citizenship is a feeling and that the status and practice of citizenship follows after this. Teachers worked to create such moments in order to quicken emotional responses as they knew this was requisite for engagement in the social issue, but also to sustain interest and agency to see the social action through. Younger students from this study were 'moved' by issues, in particular ones which seemed have an emotional connection with their lives – through places and people they knew.

It was also apparent that certain emotions and dispositions were promoted through the sociocultural norms and *sociocultural* practices of these social studies classrooms. The 'rules of the game' in these classrooms, as required by the formal national assessment, stressed that social action was an unequivocal 'good' to be pursued and at all costs. Moreover, the 'correct' or desirable emotions to demonstrate were passion, empathy and engagement, whilst cynicism and disengagement were to be avoided. Such unspoken but present notions of the 'correct' type of emotion were facilitated by unequal power relations in the classroom which serve to reinforce a certain emotional expression or habitus between teachers and students (Reay 2015, Zembylas 2005). A *sociocultural* analysis also revealed that intergenerational social interactions between young people and teachers affirmed and mediated certain types of emotion and action.

Finally it was apparent that emotions were not 'in' individuals or 'in' the classroom context, but *interacted* together to create the conditions which enabled and constrained certain emotional responses (Ahmed 2014). In this *interactionist* (Zembylas 2007) approach, student group interactions operated in different but often parallel ways to how teachers set up the classroom and illustrated the circularity of emotion. Our study exemplified how emotion can be seen as the product of social interactions that mimic wider patterns of power and control in society and in this case, the social action process had displaced the locus of responsibility management for young people's emotions from the teachers and on to some students who had assumed (or were given) leadership in their social action groups. Whilst teachers still assumed most of the control over the assessment and process, student leaders within their groups exhibited a degree of emotional labour to find ways to manage their peers. This illustrates the fluidity and pervasiveness of emotions of care (or not caring) when implementing active citizenship education policy which centred on relational connections between teachers and students and then, in turn, students and their peers – all of which were created, governed and mediated by the norms and expectations of this socio-spatial context (Zembylas 2005, 2007).

Analysing this context through Zembylas' (2007) three different approaches reveals that different theoretical approaches shape our understandings of how emotions operate in the classroom. While a *psychodynamic* approach gave insights into teachers' and students' personal feelings, *sociocultural* and *interactionist* approaches enabled us to see the relational and spatial aspects of emotions, and how these were constituted mutually and collectively between young people and adults. Moreover, the spatial context – which is this case was primarily within the context of schooling sites – shaped the nature of these interactions, serving to both constrain and enable certain forms of emotional response and citizen action. The 'politics of affect' (Thrift 2004) therefore must be necessarily understood through a socio-spatial analysis that teases out the spatiality and sociality of emotion (Ahmed 2004).

The findings of our research present some dilemmas for those involved in the context of citizenship education. While young people need to be emotionally engaged, this engagement cannot be 'forced' if their acts of citizenship are to be meaningful. The findings also suggest that amidst the busyness of school settings, time and freedom needs to be given to allow students opportunities to explore issues which they can connect and emotionally engage with.

However, when time-frames are tight, which is especially likely in the context of formally assessed social action for high stakes qualifications, more linear types of social action processes may be produced which inevitably result in some students' disengagement. The extreme of this could promote a type of social action that was functional and utilitarian, underpinned by highly 'managed' emotions, superficially producing the 'right' type of attitudes and care. This type of approach is unlikely to result in students adopting long-term social engagement and deeper understandings of citizenship through into their adult lives (Boler 1999). In contrast, there is potential for an active citizenship experience to be a reflexive and inquiring process, driven by student interest and informed by deepening knowledge about social issues and social justice and the understanding that young, caring citizens themselves can play a role in shaping society. Acknowledging how young people 'care' in citizenship contexts therefore is a vital yet complex part of the process of learning to become an active citizen.

Acknoweldgement

This work was supported by funding from the Teaching and Learning Research Initiative (2015–2016), a New Zealand government funded initiative. We are grateful for advice and comments from the editors and reviewers.

References

Ahmed, S. (2004) *The cultural politics of emotion*. London, Routledge.

Ahmed, S (2014) *The cultural politics of emotions*. Edinburgh, Edinburgh University Press.

Archer, E. and Openshaw, R. (1992) Citizenship and identity as 'official goals' in social studies. In Openshaw, R. (ed.) *New Zealand social studies: past, present and future*. Palmerston North: Dunmore Press, pp.19–33.

Bartos, A.E. (2012) Children caring for their worlds: the politics of care and childhood. *Political Geography* 31, pp.157–66.

Berg, B.L. (2007) *Qualitative research methods for the social sciences*. Boston, Pearson.

Birdwell, J., Birnie, R. and Mehan, R. (2013) *The state of the service nation: youth social action in the UK*. London, Demos.

Black, R. (2015) Between policy and a hard pedagogical place: the emotional geographies of teaching for citizenship in low socio-economic schools. *Pedagogy, Culture and Society* 23, pp.369–88.

Boler, M. (1999) *Feeling power: emotion and education*. London, Routledge.

Bondi, L. (2005) Making connections and thinking through emotions: between geography and psychotherapy. *Transactions of the Institute of British Geographers* 30, pp.433–48.

Bondi, L., Davidson, J. and Smith, M. (2005) Introduction. In Davidson, J., Bondi, L. and Smith, M. (eds.) *Emotional geographies*. Aldershot, Ashgate, pp.1–16.

Brooks, R., and Holford, J.A. (2009) Citizenship, learning and education: themes and issues. *Citizenship Studies* 13, pp.85–103.

Brown, G., and Pickerill, J. (2009) Space for emotion in the spaces of activism. *Emotion, Space and Society 2*, pp.24–35.

Creswell, J. (2014) *Research design: qalitative, quantitative and mixed methods approaches*. London, Sage.

Davies, I., Hampden Thompson, G., Tsouroufli, M., Sundram, V., Lord, P., Jeffes, J. and Bramley, G. (2012) Creating citizenship communities. *Journal of Social Science Education* 11, pp.107–18.
Farahmandpour, H. (2011) *Beyond 40 hours: meaningful community service and high school student volunteerism in Ontario*. Toronto, Ontario Institute for Studies in Education.
Freire, P. (1973) *Pedagogy of the oppressed*. London, Penguin.
Gagen, E.A. (2015) Governing emotions: citizenship, neuroscience and the education of youth. *Transactions of the Institute of British Geographers* 40, pp.140–52.
Greenberg, M.T., Weissberg, R.P., O'Brien, M.U., Zins, J.E., Fredericks, L., Resnik, H. and Elias, M.J. (2003) Enhancing school-based prevention and youth development through coordinated social, emotional, and academic learning. *American Psychologist* 58, pp.466–74.
Harris, A. and Wyn, J. (2009) Young people's politics and the micro-territories of the local. *Australian Journal of Political Science* 44, pp.327–44.
Harris, A., Wyn, J. and Younes, S. (2010) Beyond apathetic or activist youth: 'ordinary' young people and contemporary forms of participation. *Young* 18, pp.9–32.
Hochschild, A.R. (2012) *The managed heart: commercialization of human feelings*. Berkeley, Univeristy of California Press.
Holden, C., Joldoshalieva, R. and Shamatov, D. (2008) 'I would like to say that things must just get better': young citizens from England, Kyrgystan and South Africa speak out. *Citizenship Teaching and Learning* 4, pp.6–17.
Jasper, J. (1998) The emotions of protest: affective and reactive emotions in and around social movements. *Sociological Forum* 13, pp.397–424.
Jerome, L. (2012) Service learning and active citizenship in England. *Education, Citizenship and Social Justice* 7, pp.59–70.
Kennelly, J. and Dillabough, J.-A. (2008) Young people mobilizing the language of citizenship: struggles for classification and new meaning in an uncertain world. *British Journal of Sociology of Education* 29, pp.493–508.
Kennelly, J. and Llewellyn, K.R. (2011) Educating for active compliance: discursive constructions in citizenship education. *Citizenship Studies* 15, pp.897–914.
Kenway, J. and Youdell, D. (2011) The emotional geographies of education: beginning a conversation. *Emotion, Space and Society* 4, pp.131–6.
Lefebvre, H. (1991) *The production of space*. Blackwell, Oxford.
Lindquist, J. (2004) Class affects, classroom affectations: working through the paradoxes of strategic empathy. *College English* 67, pp.187–209.
Marsh, D., O'Toole, T. and Jones, S. (2007) *Young people and politics in the UK: apathy or alienation?* Basingstoke, Palgrave.
Marston, S.A. and Mitchell, K. (2004) Citizens and the state: citizenship formations in space and time. In Barnett, C. and Low, M. (eds.) *Spaces of democracy: geographical perspectives on citizenship, participation and representation*. London, Sage, pp.93–112.
Massey, D. (2004) Geographies of responsibility. *Geografiska Annaler* 86B(1), pp.5–18.
McLeod, J. (2012) Vulnerability and the neo-liberal youth citizen: a view from Australia. *Comparative Education* 48, pp.11–26.
Nairn, K. and Higgins, J. (2011) The emotional geographies of neoliberal school reforms: spaces of refuge and containment. *Emotion, Space and Society* 4, pp.180–6.
Nelson, J. and Kerr, D. (2006) *Active citizenship in INCA countries: definitions, policies, practices, and outcomes*. London: NFER/QCA.
NZQA [New Zealand Qualifications Authority] (2013) *Social studies subject resources*. Available at: www.nzqa.govt.nz/qualifications-standards/qualifications/ncea/subjects/social-studies/levels/
Osler, A. and Starkey, H. (2005) *Changing citizenship: democracy and inclusion in education*. Maidenhead, Open University Press.

Pykett, J. Saward, M., and Schaefer, A. (2010) Framing the good citizen. *British Journal of Politics and International Relations* 12, pp.523–38.

Reay, D. (2015) Habitus and the psychosocial: Bourdieu with feelings. *Cambridge Journal of Education* 45, 9–23.

Ross, A. (2012) Editorial: Education for active citizenship: practices, policies and promises. *International Journal of Progressive Education* 8, pp.7–14.

Schulz, W., Ainley, J., Fraillon, J., Kerr, D. and Losito, B. (2010) *ICCS 2009 international report: civic knowledge, attitudes, and engagement among lower-secondary students in 38 countries*. Camberwell, ICCS International Study Centre.

Sheppard, M., Katz, D. and Grosland, T. (2015) Conceptualizing emotions in social studies education. *Theory and Research in Social Education* 43, pp.147–78.

Simonsen, K. (2005) Bodies, sensations, space and time: the contribution from Henri Lefebvre. *Geografiska Annaler: Series B, Human Geography* 87, pp.1–14.

Staeheli, L.A. (2011) Political geography: where's citizenship? *Progress in Human Geography* 35, pp.393–400.

Thrift, N. (2004) Intensities of feeling: Towards a spatial politics of affect. *Geografiska Annaler* 86 B, 57–78.

Weller, S. (2007) *Teenagers' citizenship: experiences and education*. Oxon: Routledge.

Wetherell, M. (2015) Trends in the Turn to Affect: A Social Psychological Critique. *Body and Society* 21, pp.139–66.

Wood, B.E. (2013) Young people's emotional geographies of citizenship participation: spatial and relational insights. *Emotion, Space and Society* 9(1), pp.50–8.

Wood, B.E. (2015a) Excluded citizens? Participatory research with young people from a 'failing' school community. *Children's Geographies*, 14(6), pp.310–324.

Wood, B.E. (2015b) Freedom or coercion? Citizenship education policies and the politics of affect. In P. Kraftl and M. Blazek (eds.), *Children's emotions in policy and practice: mapping and making spaces of childhood*. London, Palgrave MacMillan, pp.259–73.

Wood, B.E., and Black, R. (2014) Performing citizenship Down Under: educating the active citizen. *Journal of Social Science Education: Special Issue: Civic activism, engagement and education: Issues and trends*, 13, pp.56–65.

Zembylas, M. (2005) Discursive practices, genealogies, and emotional rules: a poststructuralist view on emotion and identity in teaching. *Teaching and Teacher Education* 21, pp.935–48.

Zembylas, M. (2007) Theory and methodology in researching emotions in education. *International Journal of Research and Method in Education* 30, pp.57–72.

Zembylas, M., Charalambous, C. and Charalambous, P. (2014) The schooling of emotion and memory: analyzing emotional styles in the context of a teacher's pedagogical practices. *Teaching and Teacher Education*, 44, pp.69–80.

Zembylas, M., Charalambous, C., Charalambous, P. and Kendeou, P. (2011) Promoting peaceful coexistence in conflict-ridden Cyprus: teachers' difficulties and emotions towards a new policy initiative. *Teaching and Teacher Education* 27, pp.332–41.

6 Orphanages as spaces of care and control

Tom Disney

Introduction

Orphanages are complex institutional environments that view themselves as effective substitutes for family on a long-term basis often on a large scale, and while their exact design and size may vary across time and space they are generally understood as a residential facility which will provide care for young children until their eventual maturity (see Freundlich 2005). These spaces are often thought of as relics of the past, phenomena relegated to the Victorian era in the West, and this is often reflected in the mostly historical studies of these care spaces (see Franklin 2014, Swain 2015). For a number of reasons orphanages have ceased to be the main form of care in the UK and many western countries, increasingly seen as damaging, ineffective and costly (see Disney 2015a, Jones 1993, Lough 2003). Contemporary UK care policy now prioritises fostering or adoption for children so as to avoid placement in residential care. But in other countries orphanages persist as the dominant form of care for children without parents (Ainsworth and Thorburn 2014). Studies of contemporary orphanages remain curiously absent within geography despite these spaces demonstrating a number of notable similarities to other institutional environments such as asylums, prisons and secure care facilities which have long been staples of geographical literature (see Parr et al 2003, Philo 1987, 1989, Moran 2015). This chapter calls for the inclusion of orphanages into the field of children's geographies, which is well placed to analyse such environments.

In considering the nature of orphanages from the perspective of children's geographies, this chapter explores the relevance of orphanages to recent developments and critiques of the biosocial divide within social studies of childhood. Notably while there have been calls for biopolitics to be utilised as a means to study contemporary childhood to address the biosocial dualism (Lee and Motzkau 2011), there has been a rather limited engagement with the term both in terms of applying it to various childhood contexts, but also in critically interrogating the meaning of the term. This chapter thus provides an overview of the development of biopolitics and also considers its complicated meanings, addressing a wider call within geography for a more nuanced engagement with the term (see Rutherford and Rutherford 2013a, 2013b, Schlosser 2008). In particular this chapter situates biopolitics as one part of biopower and explores the scalar complexity of the term, which should be of interest to geographers beyond children's geographies. Finally the chapter concludes with an overview of orphanage spaces, drawing upon existing qualitative studies of these spaces to consider the operation of power across varying scales encompassing biopolitics and disciplinary power. This chapter thus addresses a need for a more critical and nuanced engagement with Foucault's theories of power, and highlights orphanages as phenomena that reflect the need within children's geographies to deconstruct biosocial dualisms and require further consideration.

Beyond biosocial dualisms

The concept of the socially constructed child has tended to drive and shape research within children's geographies, with scholars of youth striving to highlight the ways in which children as social actors negotiate their agency within environments such as the home (Holloway and Valentine 2000b, Valentine 1997), the school (Brown 2007, McGregor 2004) and public spaces (Lieberg 1995, Matthews et al. 2000). More recently however, youth scholars have begun to question the overemphasis on the social construction of the child; Prout (2005, p.2) has voiced concern that social studies of childhood are too often reinforcing dualisms childhood where "the social view of childhood is counterposed to a natural or biological one", arguing that increasingly social studies of childhood are guilty of "bracketing out or expelling biology, the body and even materiality" (ibid, p.84). Rather, he posits, childhood should be seen in hybrid terms so that elements of both are understood to be implicated in childhood, and a key question for children's scholars has been how to overcome what Lee and Motzkau (2011) term the 'bio-social dualism', and the 'wall of silence' that characterises the relationship between developmental psychology and sociological, anthropological and geographical studies of childhood (Thorne 2007). Similarly Ryan (2011), presenting a genealogy of childhood from the late 18th to the early 20th centuries, critiques the accuracy of 'bio-social dualism' as a means of theorising modern western childhood. Instead, in analysing the work of Rousseau, in particular *Émile* (1993 [1762]), he describes a "bio-social nexus" where nature and culture is understood as a "process of simultaneity" rather than a dualism or a sequence (Ryan 2011, p.446). Reflecting upon Ryan's genealogy, Pykett and Disney (2015, p.4) note,

> [it] is thus argued that a biopolitical attention to life itself has been integral to the shaping of modern childhood, making social constructivist theories of the child look partial at best and unwittingly ignorant at worst.

Thus there is a need to develop further modes or methods of understanding childhood, not to undermine the foundational principles of children's geographies and youth studies of agency and voice, but rather to allow for another element of discussion that acknowledges materiality, the body, nature and biology and thus moves beyond the 'biosocial dualism' and provides a more holistic conceptualisation of childhood.

It is notable with regards to orphanages that the biosocial dualism is equally apparent, but rather in reverse with bio-psychological studies dominating the field rather than social or cultural understandings of these environments. Our understanding of orphanage spaces has tended to be shaped by attachment theory and engagement with this through disciplines such as psychology, psychiatry, paediatrics, biology and latterly neurobiology. Attachment theory was developed by the psychiatrist John Bowlby (1969, 1973, 1980) and refined by Mary Ainsworth (Ainsworth et al 1979); Bowlby argued that from the age of 6 months to 3 years of age children form attachments with familiar adults and their development is strongly linked to these adults' (or adult's) sensitivity to their needs. More recently attachment theory has been subject to various critiques; most notably feminist scholars have revealed the ways in which the theory is imbued with normative gender assumptions about the mother's role in 'normal' child development (Burman 1995). Despite this, attachment theory has been a powerful instrument within Western child social policy and has played a key role in the deinstitutionalisation of orphan care in many countries in the West. Furthermore spaces of orphan care have increasingly become of interest to neuroscientists in attempting to illustrate the physiological and embodied impacts of institutionalisation, with images of atrophied bodies and brain scans of orphan children from such environments (Perry 2002). In contrast there has been rather less engagement with the spaces of orphan care from within the social sciences, and particularly geography, thus engendering a biosocial divide where orphanages are only understood through a particular (bio-psychological) lens.

How then do we address such divides within the orphanage literature and children's geographies more generally, without "expelling or bracketing out" biology as Prout warns (2005, p.84)? There are some (albeit limited) attempts to address this within youth studies; most relevant to this chapter is the attempt of Lee and Motzkau (2011) to move beyond the 'biosocial dualism' that characterises so many social studies of childhood. Acknowledging the diminished utility of the 'biosocial dualism' as a navigational tool for the study of childhood they draw upon Rose (2007) arguing that "life processes and social processes now appear to regularly mix with and to influence one another without regard to a biological/social boundary" (Lee and Motzkau 2011, p.8). Using the example of a teenage deterrent device termed 'Mosquito', Lee and Motzkau describe how, due to age-related hearing loss, the mosquito device's high pitched whining noise could be used to selectively target the young and disperse them from certain areas, illustrating the interplay of the natural/biological and the social. They argue that biosocial accounts of childhood are no longer adequate to explore issues where the biological intersects with the social such as childhood obesity, the influence of neuroscience upon education or the (over-)use of pharmaceuticals to treat conditions in childhood such as ADHD to name just a few. As they note "many life processes cannot be considered to lie 'outside' or 'before' social action because they themselves are a principle site of social action" (ibid, p.9). Instead their preferred tool for navigating the study of childhood is a Foucauldian approach, analysing childhood through a biopolitical lens to consider hybrid childhoods, so that the biological/natural and the social are not seen as ontologically distinct. This approach is echoed by Kraftl (2015, p.221), in particular in his exploration of what he terms 'alter-childhoods', defined as attempts to manage childhoods outside or differently from the mainstream. In the case of his research this constitutes spaces of alternative education where he describes attempts to "resist hegemonic biopower and constitute alternatives" from below, in contrast to most studies of the governance of children which takes place from above.

As Lee and Motzkau (2011, p.18) note however, this attempt to move beyond biosocial dualisms should not be seen as a 'grand unified theory'; rather this is one way in which the social-nature divide might be considered, which raises questions around what other approaches to understanding childhood beyond binary classifications there might be. For instance scholars of childhood might consider what 'natural' or even 'non-human' elements might be involved in care spaces for young people (for example see Kraftl 2013).

Biopolitics and biopower are effective tools to understand orphanages in particular, acknowledging these institutions as spaces of care, health and embodiment but also as socially and culturally constructed environments as well. Before a review of the nature of orphanages as spaces of care and control, it is necessary to discuss and consider the complexities of biopolitics.

Considering biopolitics

While there has been a marked proliferation of the use of the term biopolitics in geography, as Rutherford and Rutherford (2013a, 2013b) note the term can have different and at times contested meanings. As they argue, the differentiated forms of biopolitics that have emerged are indicative of the increasing usage and evolution of the term, including 'necropolitics' (Mbembe 2003, Davies and Isakjee 2015), 'thanatopolitcs' (Joronen 2015, Tyner 2015), and 'ethnopolitics' (Rose 2007) among others. Rutherford and Rutherford (2013a, 2013b) highlight a concern that at times the term is being adopted uncritically within geography. Within children's geographies in particular there has tended to be a rather limited engagement with biopolitics; despite Lee and Motzkau (2011) advocating the use of biopolitics to analyse childhood few children's geographers provide any detailed examination of the term and often devote little more than a cursory mention to it. Kraftl (2015, p.220–2) similarly notes the lack of engagement with the term, and his excellent piece on 'alter-childhoods' represents a more critical analysis of the term in which he draws upon the work of Giorgio Agamben in considering alternative spaces of education for children in the UK. With the confusion over the term noted by Rutherford and Rutherford (2013a, 2013b) and the lack of critical engagement with biopolitics by children's geographers, this chapter is concerned with the origins of the term, which lie largely with Michel Foucault, and how orphanages can be read through his understandings of power.

In a broad sense, a major aspect of Foucault's work can be seen as describing the development of different forms of power, most notably the shift from sovereign power (such as that of a monarch) to power "focused on the governance of people" or biopower (Schlosser 2008, p.1622). Although often mentioned in conjunction with biopolitics, biopower is not quite the same as biopolitics; rather biopolitics is encompassed within biopower and only one element of it, alongside other forms of power such as disciplinary power for example.

In describing the shift to modern biopower (the power over *bios* or life) and the state's growing interest in the health of its citizens, Foucault (1998) notes that:

> Power would no longer be dealing simply with legal subjects over whom the ultimate dominion was death, but with living beings, and the mastery it would be able to exercise over them would have to be applied at the level of life itself: *it was the taking charge of life, more than the threat of death, that gave power its access even to the body*. (pp.142–3, emphasis mine)

According to Foucault (ibid, p.139), biopower emerged in two basic forms, which he describes as 'two poles' interlinked through a variety of different relations:

> [The first] centred on the body as a machine: its disciplining, the optimisation of its capabilities, the extortion of its forces, the parallel increase of its usefulness and of its docility, its integration into systems of efficient and economic controls, all this was ensured by the procedures of power that characterised the *disciplines*: an *anatomo-politics of the human body*. The second, formed somewhat later, focussed on the species body, the body imbued with the mechanics of life and serving as the basis of the biological processes: propagation, births and mortality, the level of health, life expectancy and longevity, with all the conditions that can cause these to vary. Their supervision was effected through an entire series of interventions and *regulatory controls: a bio-politics of the population*. (p.139, emphasis in original

As Rutherford and Rutherford (2013a, p.414) note, what "became so important to governance was *the norm*, indeed the elaboration of the normal and the abnormal, rather than law, which was the mark of sovereignty" (emphasis in the original). The aim of contemporary biopolitics and biopower is today less often the eradication and segregation of those deemed abnormal, and more about the correction, management and treatment of populations (Rose 2001).

There is also a scalar complexity to biopower, and it is focussed not only at the level of populations but also targets individual bodies, which are necessarily intertwined as populations are made up of individual bodies. As Taylor notes (2011, pp.44–5) at one level there are institutions such as psychiatric hospitals, asylums, prisons, schools and arguably orphanages which focus disciplinary technologies on individual bodies to prevent them from deviating from the norm, and then there is the macro-scale, societal level of state biopolitics interested in the health and 'normality' of the population as a whole.

Schlosser (2008, p.1624) makes an important point with regards to Foucault's conceptualisations of power, often seen as a progression or linear development; although there are different forms of power, he does not necessarily see these as acting in opposition to each other, but rather operating with a 'field of power'. Indeed as Foucault himself (1991b) notes:

> [We] need to see things not in terms of the replacement of a society of sovereignty by a disciplinary society and the subsequent replacement of a disciplinary society by a society of government; in reality one has a triangle, sovereignty-discipline-government, which has as its primary target population as its essential mechanism the apparatuses of security. (p.102)

With this in mind, Foucault's conceptualisations of power look less like a linear 'progression' and more lie a 'triangle' or 'triad' of different forms of power with emphasis on certain types of power at certain times or in certain contexts. Importantly then for this chapter, while Foucault does describe a general shift away from sovereign power towards more subtle forms of disciplinary power, he does not necessarily suggest that other forms of power do not continue to exist along multiple scales. A more nuanced reading of biopolitics should take account of the multi-scalar nature of biopower and its multifaceted aspects, inclusive of different forms of power, both that which is creative and positive, and that which is encompasses severe forms of discipline and punishment. With regards to care then questions should be asked of the wider discourses that surround institutions of health and care for young people, the macro-scale interventions that lead to children entering these spaces and why relations within these environments might at certain times be positive and therapeutic and at other times violent and coercive. What is it that leads care into control? When does the relationship between these two poles become uneven?

This chapter now moves to consider orphanages as ostensibly spaces of care, but which can also exhibit severe elements of control as well; indeed they provide a powerful example of the intersection of these different forms of power across different scales. With this in mind this chapter now considers how orphanages demonstrate these different forms and scales of power through a consideration of the macro-level movement into these spaces, and the elements of control that take place within them.

Biopolitical orphans

While often greatly varying across geographical borders and contexts, orphanages are all technically designed as environments of care that aim to provide for the health and wellbeing of children without parental care (see Freundlich 2005 for more on orphanage design). Such practices reflect foundational concerns within children's geographies and youth studies in which children are conceptualised as 'at risk' or constituting 'a risk' (Holloway and Valentine 2000a), with the state moving to intervene in 'at risk' children's lives in the name of health and care. As Fonseca (1986, p.15) notes in her research exploring orphanages, foundlings and foster mothers in Brazil, the placement of children in orphanages or residential care spaces "occur under 'crisis conditions' – when, the child's home has been ruptured by death, divorce or remarriage". Similarly McIntosh et al. (2010, p.296) note that children often enter such spaces following a breakdown of parental care and supervision, or because of neglect of the health and care of the child. In such instances the health and care of the young population is the fundamental concern, revealing the macro-scale biopolitical state processes.

Children can also enter such spaces because of a perception that these environments offer certain advantages that are unavailable elsewhere; Carpenter (2015, p.85) notes that "many Cambodians are positively inclined towards orphanages because of educational opportunities they see orphanages providing". Such findings are similarly reported by Emond (2009, p.413), who found that some orphans considered themselves 'lucky' to have been saved from abuse and to be receiving educational opportunities (see also Emond 2010). Entry into these spaces can be a form of care in itself; when children are provided with the choice of whether they wish to live in such an environment it can be a form of empowerment and a way to facilitate agency (see Disney 2015a). These studies highlight the heterogeneity of orphanage environments, with these particular spaces and practices reflecting the more benign biopolitical power relations that can draw agents into orphanages, and the fact that such power is not inherently negative or positive necessarily. These forms of movement into orphan environments for the purposes of care, education, health and child protection can be read as a form of biopolitical power aimed at the maintenance of a healthy population.

While orphanages can exhibit these more benign forms of power, these institutions also have the capacity to be emblematic of the darker side of biopolitics, with governments targeting 'risky', unhealthy or 'defective' elements of society. Drawing upon the narratives of former inmates of the Parramatta Girls Home in Australia, and submissions to public enquiries, Franklin (2014) explores women's relationships to their former residence, in particular their ability to create a sense of place among violence and abuse. Notably young girls entered this institution having been identified by state services as 'dangerous' and 'risky', and thus in need of segregation from the 'normal' mainstream population: "[from] 1887 to 1974, 'corrupt' adolescent girls were incarcerated in Parramatta Girls Home, a New South Wales state government facility charged with their 'care and protection'" (Franklin 2014, p.158). The ways in which these young women were drawn into these spaces highlights the complexity of state biopolitics, where care and control intersect or are not necessarily distinct. Indeed as Franklin notes of those entering the institution, there was as an eventual blurring of 'the delinquent' and 'the destitute' youth identified as requiring institutionalisation in the home.

These forms of biopolitics aimed at the institutionalisation of risky or unhealthy youth elements into orphanages persist in contemporary contexts; in Russia orphans are a specific target for the state's biopolitical control on the macro scale, their importance underscored by a continuing demographic crisis in the country (Eberstadt 2011), and the perception that the state can provide better care within institutions for certain children such as those termed at risk from 'unstable' families (Khlinovskaya Rockhill 2004, 2010). Khlinovskaya Rockhill highlights a cycle of child displacement in the Russian north, where families who are termed *neblagopoluchnyi* (translated roughly 'problem family', see Khlinkovskaya Rockhill 2004 for further details) are targeted by the authorities move to deprive the parent(s) of their rights to the child. The identification of these parents as *neblagopoluchnyi* and the decision to remove the children are based on "two contradictory concepts of the person's moral nature … : 'environmental' and 'genetic' (the concept of 'bad genes'). Many families are viewed by the authorities as 'unfit', owing to moral judgements underpinned by Soviet values" (Khlinovskaya Rockhill 2004, p.147). Ironically it appears that having spent time in the *detdom* (state orphanage) is one element involved in the development of a person who can later become classified as *neblagopoluchnyi*, so the state intervenes to remove the child from a *neblagopoluchnyi* family only to facilitate the child's own development into a person which the state views as a threat (to future children). The processes at work here are emblematic of the problematic nature of these biopolitical interventions however; they are ostensibly in the name of health and yet at times they are (in the eyes of the state) producing tainted and problematic citizens that are again targets of biopolitical concern.

The identification of 'defects' in some form can also lead to institutionalisation in Russia (see Klepikova and Utekhin 2010). In Post-Soviet Russia the lines between orphanage and disability institution are porous and complex. Some children enter the system orphanage system after being legally 'refused' by their parents sometimes under pressure from the medical establishment following the identification of a disability, or because they are considered to require significant medical treatment (Klepikova and Utekhin 2010). The orphanage system in Russia divides at the point that children leave the baby home (a space for younger at risk or rejected children), at age three or four, when they are assessed by the '*Psikhologo-mediko-pedagogicheskaya Komissiya*' or PMPC (Psychological-Medical-Pedagogical Commission), a committee which includes a pedagogue, psychiatrist, psychologist and speech therapist who evaluate "the children's development and assigns them either to a pre-school Children's Home for normal children or a Children's Home for disabled children" (Khlinovskaya Rockhill 2010, p.198) An assessment that a child is 'disabled' leads into a complex web of institutional environments, an orphanage system for children with impairment numbers 1–8 with each category of institution designed to theoretically offer support to the specific impairment or needs of the child, with the PMPC operating a sorting process to categorise children and place them in the relevant institutions that will theoretically meet their specific needs (Disney 2015b, Human Rights Watch 2014). Such practises highlight the complexity of biopolitics in different contexts, and the capacity for the darker side of these forms of power to persist.

Orphanages as spaces of care and control

On the micro scale orphanages can also be seen to encompass other elements and scales of biopower, particularly disciplinary power, highlighting that biopolitics does not necessarily operate in isolation from or in opposition to other forms of power. It is notable that Foucault (1991) saw the completion of the carceral regime not in the Panopticon (so often seen as *the* disciplinary model in geographical and sociological studies of Foucault (see Philo 2011)), but in Mettray, the colony for juvenile delinquents and home to many orphans in France. For Foucault (1991a, p.293) Mettray represented the "disciplinary form at its most extreme, the model in which are concentrated all the coercive technologies of behaviour".

Mettray was known as "the 'prison' without walls" (Driver 1990, p.276) because the 'moral training' the children received meant that they would not try to escape. Frederic Demetz, the founder of Mettray, hoped to instil in each inhabitant a moral seed so that he would reform himself and thus self-discipline and no longer be so vulnerable to social ills and criminal behaviour (Driver 1990). The use of families in Mettray was integral to instilling this moral seed and critically *loyalty* to the institution so that the each child became an agent of surveillance (Philo 2011). Mettray's success seemed conclusive when in 1848 there were uprisings in a number of prisons across France, yet Mettray and its inmates remained calm and untroubled (Dekker 2007).

Mettray has long since disappeared, but the nature of this reformatory for children as a severe disciplinary institution importantly marks spaces of orphan care as environments that should be considered alongside other disciplinary institutions such as prisons and asylums, not least because they all contain a population mostly segregated from the rest of society often (but not always) against this population's own will or desire. This highlights the complexity of the orphanage as a disciplinary care space; it is designed to care for children who are both at risk and constitute a potential future risk to society either in terms of their delinquent status or because they threaten the maintenance of the health of the population. Surveillance, discipline and punishment within orphanage environments are conducted theoretically with the aim of care; at times, however, these can be warped and abusive. Thus orphanages as environments of care can be understood as institutions that also punish, discipline and control. Given the heterogeneity of orphanage environments, power also exists differently in these environments with some exhibiting elements of discipline and control that are particularly severe on the micro scale, alongside macro-scale practices. This final section again draws upon qualitative studies of orphanage environments to highlight the lived experience of orphanage environments and illustrate the ways in which such spaces cannot be understood through a purely bio-psychological lens and thus the limitations of the biosocial dualism in such environments. Finally elements of control, discipline and punishment are noted in these spaces and thus their similarity to other institutional environments such as prisons and asylums.

Control, discipline and surveillance

Drawing upon recently released documents detailing the experiences of graduates of state institutional care in Australia, Swain (2015) notes how despite appearance orphanage buildings were often a means of segregating a group of children away from the rest of the population:

> [Whether] they occupied formerly grand mansions or new structures built for purposes, such grandiose exteriors often disguised an inner penury where children were contained and controlled. (p. 18)

Franklin (2014) similarly describes the way in which the Parramatta Girls home was designed to contain and control:

> Control was maintained in many ways. The building and grounds were physical indicators that the 'Home' was not a home, not a place of security and belonging, but indeed a prison-like institution. (p.160)

Confirming the prison-like nature of the institution were the reflections of former residents of the home who described how the perimeter walls were designed to prevent the girls escaping, rather than anyone getting in: "[the wall is] cemented out on a 45, 35 degree angle and there's all broken glass cemented in it."

Swain (2015, pp.19–20) notes the regimentation, functionality and surveillance of life within these spaces, in particular in relation to hygiene in which one former resident recalls that the "system was as efficient as a factory assembly line: six shower outlets to each of five rows, thirty boys at a time". The regimentation and militaristic organisation is demonstrative of the disciplinary power of such spaces, and echoes Foucault's (1991a) description of 'docile bodies', whereby individuals are moulded into productive beings. As Hoffman (2011, p.28) points out such disciplinary power attempts

> to meticulously, exhaustively and continuously control the activities of bodies so as to constitute them as bearers of a highly particular relationship between utility and docility, whereby increases in utility correspond to increases in docility and vice versa.

Within these care spaces disciplinary power is enacted to mould the citizens of the future, and engender docility in these orphan bodies in order to maximise later economic utility.

Alongside elements of regimentation and surveillance, Swain (ibid., p.24) also highlights practices of constraint and containment noting that "[offenders] and absconders found themselves locked in tiny cells inside barbed wire-topped walls where the model appeared to be the jail rather than a home". Similar practices of constraint and containment have been noted in orphanages in other more contemporary contexts too; Human Rights Watch (2014, p. 43) found forms of constraint within Russian orphanages in their recent report:

> In a Moscow specialized infant care institution, one staff member [revealed] that she had bound a 4-year-old blind girl to a wheelchair using rags tied around the girl's head and torso, despite the fact that the girl could walk.

Further practices of constraint have been noted in Russian orphanages such as medical and psychological forms of restraint and control (see Disney 2015b).

Colton (2002, p.37) describes a process of dehumanisation in discussing abuse in UK state institutional care, and attempts to explain how violence is able to take place with these institutions suggesting that victims "are seen as less than fully human by their oppressors who, in the absence of effective systems of internal and external monitoring, wield almost absolute power". The notion of dehumanised identities echoes Goffman's (1961) 'total institutions' and suggests that orphanages might be included in the list of such institutions. While much of the current bio-psychological literature highlights the ways in which children might be physiologically deprived by these environments (see O'Connor et al 2000), this is demonstrative of the ways in which such spaces are able to shape individuals through social processes as well, demonstrative of the inadequacy of biosocial dualistic approaches to orphanages.

Ultimately this is reflective of imbalanced power relations within these spaces of care, reflecting Foucault's (2008) explorations of power in asylums that were characterised by a deeply unequal power relationship between the patient and psychiatrist in the asylum. Similarly, alongside the aforementioned practices of constraint and containment, it is indicative of the perpetuation of more severe forms of power and is illustrative of Foucault's 'triangle' of various forms of power at work.

While these institutional environments can encompass these warped forms of care, they are also highly heterogeneous and engender complex emotional responses; Franklin (2014) notes that many former inhabitants formed a strong sense of place and attachment to the former institution, with some positive memories despite the horror of abuse, suggesting that children are able to form a sense of place at times in these complex environments. Such findings are reflected in studies of asylum spaces; for example Parr et al (2003) found that patients in the former asylum of Craig Dunain similarly expressed a sense of place with reference to the asylum, and particularly its 'free places' (see Goffman 1961). Research into the geography of prisons has also revealed such complex responses from inhabitants; for instance Crewe et al (2014) note that

> [despite] ostensible claims that emotion and intimacy were absent from prison life … many prisoners in fact exposed their vulnerabilities, albeit in controlled forms and within limited circles, soliciting advice on personal relationships and seeking kindness and support from trusted associates. (p.66)

The similarities with the findings of asylum and carceral geographies highlight potential avenues of research that children's geographers might contribute to, in particular through an exploration of orphanage spaces and the importance of place in environments of care and control. The ability to create a sense of place in such institutions might also be read along Foucauldian lines, with inmates, patients and orphans in such spaces internalising disciplinary power practices, much in the way the Foucault (1991a) described the loyalty of the children in Mettray. The complex emotional engagements with these environments also highlights the importance of analysing orphanages beyond the bio-psychological lens, as they encompass different practices, processes and emotions which the biological alone cannot illustrate.

The heterogeneity of such environments is reflected in the aforementioned work of Carpenter (2015) and Emond (2010) where collective living and educational opportunities in Cambodian orphanages meant that many children reported positive experiences of such environments. The complexity of these environments, which can encompass severe forms of discipline and punishment for some but also positive associations of place for others is reflective of the complex nature of power which is can be similarly experienced in various forms. It is notable that Foucault (1991a, p.303) also saw a relationship between care and control for the purposes of discipline: "in its function, the power to punish is not necessarily different from that of curing or educating".

Conclusion

Recent developments in children's geographies and youth studies have highlighted the limitations of the dualistic biosocial approach to the study of childhood, whereby biology is expelled or bracketed out (Prout 2005). Orphanages reflect these concerns, albeit in reverse, with bio-psychological studies shaping our current understandings of spaces of orphan care and a lack of socio-cultural understandings of these environments. As a way forward some have advocated the use of biopolitics (Lee and Motzkau 2011), and yet there has been a rather limited engagement with the term and more broadly within geography there are concerns that the term is being used uncritically (Rutherford and Rutherford 2013a, 2013b, Schlosser 2008). This chapter has explored the origins of the term, which lie largely with Michel Foucault. Of particular importance is the understanding that biopolitics represents one facet of wider aspects of power that Foucault describes, in particular biopower, which also encompasses other forms of power such as disciplinary power. As discussed, many see Foucault's description of power as a linear development, or progression from violent sovereign power to subtle, surveillance-based disciplinary power, when it appears as though he sees a 'triangle'/'triad' of different forms of power, with an emphasis on certain types of power at different times or within certain contexts. Such a conceptualisation of power suggests biopolitics should not be considered in isolation from other forms of power, and that the scalar complexity of biopower requires further exploration from geographers and scholars of childhood.

Orphanages represent a powerful and under-researched example of these different forms and scales of power in operation; on the macro-scale state level biopolitical concerns around 'at risk' children are evident, as certain children are moved from 'crisis' environments into spaces of orphan care, which are thought to provide more effectively for the health and care of these children. Alongside these seemingly benign forms of biopolitics, there are also forms of biopolitics that target the 'risky', 'unhealthy' and 'defective' inclusive of perceived delinquents and disabled children, who are drawn into these spaces as they are seen to endanger the health of the population. On the micro scale, within orphanage spaces, the other elements of biopower and Foucault's 'triangle' of power are evident, with elements of disciplinary power that can at times be severe and constraining as well as elements that can be more benign. The complexities of these power relations highlight the inadequacy of the biosocial dualism for the study of orphanages as socially and culturally constructed environments as well and experienced in embodied terms. This application of Foucauldian notions of power to these spaces also highlights the relevance of such environments to geography, and this chapter calls for consideration of what a spatial analysis of orphanages might offer alongside prisons, secure care facilities and other institutional spaces of childhood.

The various forms of power operating in institutions of care for young people require further attention from scholars of childhood; in particular the relationship between care and control is poorly understood. At what point does the relationship between care and control become uneven, unstable perhaps? What takes place to destabilise this relationship and allow violence and severe discipline to become a part of these spaces? Further research is required into the various everyday spaces of care that young people experience, such as orphanages or other such institutional environments, in order to further understand this relationship more fully in various contexts.

Acknowledgement

This work was supported by the Economic and Social Research Council (grant number ES/J50001X/1).

References

Ainsworth, F. and Thoburn, J. (2014) An exploration of the differential usage of residential childcare across national boundaries. *International Journal of Social Welfare* 23, pp.16–24.

Ainsworth, M., Blehar, M., Waters, E. and Wall, S. (1979) *Patterns of attachment*. Hillsdale, Elbaum.

Bowlby, J. (1969) *Attachment and loss vol. 1: attachment*. London, Hogarth Press.

Bowlby, J. (1973) *Attachment and loss vol. 2: separation, anxiety and anger*. London, Hogarth Press.

Bowlby, J. (1980) *Attachment and loss vol. 3: loss, sadness and depression*. London: Penguin.

Brown, J. (2007) Time, space and gender: understanding 'problem' behaviour in young children. *Children and Society* 21, pp.98–110.

Burman, E. (1995) The abnormal distribution of development: policies for Southern women and children. *Gender, Place and Culture* 2, pp.21–36.

Carpenter, K. (2015) Continuity, complexity and reciprocity in a Cambodian orphanage. *Children and Society* 29, pp.85–94.

Colton, M. (2002) Factors associated with abuse in residential child care institutions. *Children and Society* 16, pp.33–44.

Crewe, B., Warr, J., Bennett, P. and Smith, A. (2014) The emotional geography of prison life. *Theoretical Criminology* 18, pp.56–74.

Davies, T. and Isakjee, A. (2015) Geography, migration and abandonment in the Calais refugee camp. *Political Geography* 49, pp.93–5.

Dekker, J. (2007) Philanthropic networks for children at risk in nineteenth-century Europe. *Pedagogica Historica* 43, pp.235–44.

Disney, T. (2015a) Complex spaces of orphan care – a Russian therapeutic children's community. *Children's Geographies* 13, pp.30–43.

Disney, T. (2015b) The role of emotion in institutional spaces of Russian orphan care: policy and practical matters. In Blazek, M. and Kraftl, P. (eds.) *Children's emotions in policy and practice: mapping and making spaces of childhood*. Basingstoke, Palgrave, pp.17–33.

Driver, F. (1990) Discipline without frontiers? Representations of the Mettray reformatory colony in Britain, 1840–1880. *Journal of Historical Sociology* 3 pp.272–93.

Eberstadt, N. (2011) The dying bear: Russia's demographic crisis. *Foreign Affairs* 90, pp.95–108.

Emond, R. (2009) 'I am all about the future world': Cambodian children's views on their status as orphans *Children and Society* 23 pp.407–17.

Emond, R. (2010) Caring as a moral, practical and powerful endeavour: peer care in a Cambodian orphanage. *British Journal of Social Work* 40, pp.63–81.

Fonseca, C. (1986) Orphanages, foundlings, and foster mothers: the system of child circulation in a Brazilian squatter settlement. *Anthropological Quarterly* 59, pp.15–27.

Foucault, M. (1991a) *Discipline and punish – the birth of the prison*. London, Penguin.

Foucault, M. (1991b) Governmentality. In Burchell, G., Gordon, C. and Miller, P. (eds.) *The Foucault effect: studies in governmentality*. Chicago, University of Chicago Press, pp.87–104.

Foucault, M. (1998) *History of sexuality Volume I: the will to knowledge*. London, Penguin.

Foucault, M. (2008) *Psychiatric power: lectures at the College de France, 1973–1974*. Basingstoke, Palgrave.

Franklin, C. (2014) Belonging to bad: ambiguity, Parramatta girls and the Parramatta Girls Home. *Geographical Research* 52, pp.157–67.

Freundlich, M. (2005) A return to orphanages? *Adoption Quarterly* 9, pp.1–11.

Goffman, E. (1961) *Asylums – Essays on the social situation of mental patients and other inmates*. New York, Anchor.

Hoffman, M. (2011) Disciplinary power. In Taylor, D. (ed.) *Michel Foucault*. Durham, Acumen, pp.27–41.

Holloway, S. and Valentine, G. (2000a) Children's geographies and the new social studies of childhood. In Holloway, S. and Valentine, G. (eds.) *Children's geographies*. London, Routledge, pp.1–28.

Holloway, S. and Valentine, G. (2000b) Spatiality and the new social studies of childhood. *Sociology* 34, pp.763–83.

Human Rights Watch (2014) Abandoned by the state: violence, neglect, and isolation for children with disabilities in Russian orphanages. Available at: www.hrw.org/sites/default/files/reports/russia0914_ForUpload.pdf

Jones, M. (1993) Decline of the American orphanage, 1941–1980. *Social Service Review* 67, pp.459–80.

Joronen, M. (2015) 'Death comes knocking on the roof': thanatopolitics of ethical killing during Operation Protective Edge in Gaza. *Antipode* online early.

Klepikova, A. and Utekhin, I. (2010) Rebenok s «otkoleniyami razvitiya»: opyt analiza freimov. *Antropologicheskii forum* 12, pp.1–68.

Khlinovskaya Rockhill, E. (2010) *Lost to the state: family discontinuity, social orphanhood and residential care in the Russian Far East*. Oxford: Berghahn Books.

Khlinovskaya Rockhill, E. (2004) Social orphans and the *neblagopoluchnaia* family: the cycle of child displacement in the Russian north. *Sibirica* 4, pp.132–49.

Kraftl, P. (2013). Beyond 'voice', beyond 'agency', beyond 'politics'? Hybrid childhoods and some critical reflections on children's emotional geographies. *Emotion, Space and Society* 9, pp.13–23.

Kraftl, P. (2015) Alter-childhoods: biopolitics and childhoods in alternative education spaces. *Annals of the Association of American Geographers* 105, pp.219–37.

Lee, N. and Motzkau, J. (2011) Navigating the bio-politics of childhood. *Childhood* 18, pp.7–19.

Leiberg, M. (1995) Teenagers and public space. Communication Research 22(6), pp.720–740.

Lough, B. (2003) The rise and demise of orphanages in Ukraine. *European Journal of Social Work* 6, pp.49–63.

Matthews H., Limb M. and Taylor M. (2000) The street as thirdspace: class, gender and public space. In Holloway, S. and Valentine, G. (eds.) *Children's geographies: Living, playing, learning and transforming everyday worlds*. London, Routledge, pp. 63–79.

Mbembe, A. (2003) Necropolitics. *Public Culture* 15, pp.11–40.

McGregor, J. (2004) Spatiality and the place of the material in schools. *Pedagogy, Culture and Society* 12(3), 347–372.

McIntosh, I., Punch, S., Dorrer, N. and Emond, R. (2010) 'You don't have to be watched to make your toast': surveillance and food practices within residential care. *Surveillance & Society* 7, pp.287– 300.

Moran, D. (2015) *Carceral geography – spaces and practices of incarceration*. Ashgate, Farnham.

O'Connor, T., M. Rutter, N., Beckit, C., Keaveney, L. and Kreppner, J. (2000) The effects of global severe privation on cognitive competence: extension and longitudinal follow-up. *Child Development* 71, pp.376–90

Parr, H., Philo, C. and Burns, V. (2003) 'That awful place was home': reflections on the contested meanings of Craig Dunain asylum. *Scottish Geographical Journal* 119, pp.341–60.

Perry, B. (2002) Childhood experience and the expression of genetic potential: what childhood neglect tells us about nature and nurture. *Brain and Mind* 3, pp.79–100.

Philo, C. (1987) 'Fitl for an asylum': the historical geography of the nineteenth-century 'Mad-Business' in England as viewed through the pages of the *Asylum Journal*. *Journal of Historical Geography* 13, pp.398–415.

Philo, C. (1989) 'Enough to drive one mad': the organisation of space in 19th-century lunatic asylums. In Wolch, J. and Dear, M. (eds.) *The power of geography – how territory shapes social life*. Winchester, Unwin Hyman, pp.258–90.

Philo, C. (2011) Foucault's children. In Holt, L. (ed.) *Geographies of children, youth and families: an international perspective*. Routledge, London, pp.27–54.

Prout, A. (2005) *The future of childhood*. London, Routledge.

Pykett, J. and Disney, T. (2015) Brain-targeted teaching and the biopolitical child. In Kallio, K.P. and Mills, S. (eds.) (2015) *Politics, citizenship and rights*, vol. 7 of Skelton, T. (ed.) *Geographies of children and young people*. Springer: Singapore.

Rose, N. (2001) The politics of life itself. *Theory, Culture and Society* 18, pp.1–30.

Rose, N. (2007) *The politics of life itself: bio-medicine, power and subjectivity in the 21st century*. Princeton, Princeton University Press.

Rousseau, J.J. (1993 [1762]) *Emile*. London, JM Dent, Everyman.

Rutherford, P. and Rutherford, S. (2013a) The confusions and exuberances of biopolitics. *Geography Compass* 7, pp.412–22.

Rutherford, S. and Rutherford, P. (2013b) Geography and biopolitics. *Geography Compass* 7, pp.423–34.

Ryan, K. (2011) The new wave of childhood studies: breaking the grip of bio-social dualism? *Childhood* 19, pp.439–52.

Schlosser, K. (2008) Bio-political geographies. *Geography Compass* 2, pp.1621–34.

Swain, S. (2015) Institutionalised childhood: the orphanage remembered. *Journal of the History of Childhood* 8, pp.17–33.

Taylor, C. (2011) Biopower. In Taylor, D. (ed.) *Michel Foucault*. Durham, Acumen, pp.41–54.

Thorne, B. (2007) Editorial: crafting the interdisciplinary field of childhood studies. *Childhood* 14, pp. 147–52.

Tyner, J.A. (2015) Population geography II: mortality, premature death, and the ordering of life. *Progress in Human Geography* 39, pp.360–73.

Valentine, G. (1997) 'My son's a bit dizzy.' 'My wife's a bit soft.': Gender, children and cultures of parenting. *Gender, Place and Culture* 4(1), pp.37–62.

7 Carefully controlled

Young people and their pathways through spaces of secure care

Anna Schliehe and Annie Crowley

Geographies of care and control

Through the prism of secure care institutions, geographies of care and control are explored in order to understand the complex and opaque nature of closed spaces in this most secure form of child and adolescent care in Scotland. These environments are used as a 'last resort' for young people who are deemed to pose a danger either to others or themselves. They provide a wide range of care while at the same time ensuring high security standards, surveillance and control of behaviour. As purpose-built secure and secluded spaces these units are intrinsically linked with geographies of children and young people as well as health and wellbeing. The exhibited geographies of care and control are a relatively new field within geographical studies on young people which is why it is helpful to consider carceral geographic and criminological literatures to reveal the complex nuances and power relations at work in secure care units.

As the most secure form of residential care available, these units provide an important albeit challenging research environment that is situated in wider debates around increasing securitisation of childhoods and criminalisation of young people (see Valentine 1995, 2003, Brown 2005). The disruption of the moral order by 'dangerous children' (Valentine 1995, p.597) is summarised as a process of marginalisation that takes specific shape in secure care. In an attempt to situate young teenagers in geographies of children and youth, Weller (2006, p.104) focuses on the labels of being 'at risk', 'in trouble' or 'in need' that are so readily applied to teenagers, which shows how this group of young adults is often constructed through negative imagery. This applies particularly to young people who find themselves incarcerated, and within this group, particularly to young women who additionally may be viewed as having transgressed traditional gender roles (Burman and Batchelor 2009). Brown (2005) conceptualises young people who are deemed 'unmanageable' as a marginal societal group she calls 'non-persons' due to binary and exclusionary views on young people in the criminal justice system. While the term 'youth' can be seen as bridge between childhood and adulthood, Barry (2006, p.25) describes it as more than a liminal phase as it involves several transitional periods. Many young people experience discrimination in social, legal and economic terms as a result of their age, and additionally, with particular relevance for this chapter, because of gender. Brown (2005) raises a similar point about the exclusion of youth underlining the marginal position of the young women in this chapter. These discussions on youth and incarceration raise contradictory issues around care/welfare and control/punishment, but also wider marginalisation and discrimination.

While there is a comprehensive range of literature on historic and contemporary effects of securitisation of youth (see Collins and Kearns 2001, Valentine 2003, Brown 2005), much less is known of the institutional geographies and pathways into locked environments for children and young people. In policy terms, extensive debates have been fought over the inherent paradox of 'care versus control' or 'welfare versus punishment' (Harris and Timms 1993, Bullock et al 1998, Goldson 2002). While some studies emphasise the punitive character of secure units, others highlight the comprehensive care and pro-social and therapeutic role modelling available (Rose 2002). There are a range of options from which public policy is formulated, notably, as indicated above, the stark choice between punishment and welfare. This is the so-called 'deep' end of the criminal justice system, where punishment and welfare are deployed as alternative strategies in what is called the 'penal-welfare complex' (Bond-Taylor and Jones 2009).

In Scotland, both the political attention towards – and the work practice to address – youth and female detention have expanded greatly since Devolution in the mid-1990s. Previous initiatives like the Kilbrandon Report (see Scottish Government 1995) also need to be taken into account when assessing the policy landscape for youth justice. For a long time Scotland stood in contrast to other jurisdictions in its penal-welfare ethos in youth justice. Since the mid-1990s, however, Scotland's adherence to penal-welfarism has steadily eroded and is shifting further away from a concern with the social and personal needs of young people to a greater focus on the nature, extent, frequency and improved surveillance of offences (Burman and Batchelor 2009, p.272). Secure care units, rather than emphasising one alternative over the other, incorporate both in equal measure and thus provide a unique environment of intertwining care and control: both merge into the other and become indistinguishable. With its distinct amalgamation of control and care, this most secure form of residential accommodation raises the need for further research into the institutional complexity and ambiguity to explore questions around effects of such high levels of control and care.

The management of 'risk' is one of the main tasks deemed as necessary in order to achieve a manageable and controllable atmosphere in the units (Roesch-Marsh 2014). In the young people's pathways and the stories they tell, it becomes apparent how the terms of security, control and risk management feature in many care arrangements, although being at their most pronounced in secure care (Schliehe 2015). In this chapter we aim to look at young people's entangled pathways in and out of secure units in order to better understand how they experience their institutional journeys, and how they understand why they are contained in such secure and locked-up conditions. We also attempt to offer professional voices to highlight some of the dilemmas and ambiguities of providing care while exercising these high levels of security.

The chapter revolves around our empirical doctoral findings featuring descriptions of secure spaces and their institutional constitution as well as offering a more in-depth understanding of the entangled pathways of young people detained in secure care. The adolescents' own experiences are described as 'moving stories' that contain interview material reflecting their journeys. The 'frontline' perspectives of staff working with young people are examined as an important aspect of understanding these journeys, as their practical application of policy means that they are themselves, in a sense, policy makers. This empirical material is embedded in an overview of secure care in Scotland examining other studies on closed environments for children and young people. Following this, the research methodologies and key features of our research are outlined. The fifth section finally draws together the main arguments and experiences of secure care and its way of carefully controlling young people and their pathways.

Secure care in Scotland

Of the thousands of children and young people who are in contact with social workers, and 'looked after' in care placements outside of their homes, only a few are placed in locked, 'secure' conditions (Harris and Timms 1993). For example, on 31 July, 2015, there were 15,404 looked after children in Scotland, with an average of just 82 residents in secure accommodation throughout 2014–15 (Scottish Government 2016). Secure care therefore represents a last resort but is nonetheless an important part of the available care establishments in its actual practice of care and control, as well as in its symbolic effect on a larger group of young people. The Secure Accommodation (Scotland) Regulations describe secure care as "for the purpose of restricting the liberty of children in a residential establishment" (SASR 2013, 2[1]), with three justifications given for securing young people (see Table 7.1).

Recommendations to place young people (up to the age of 18) in secure accommodation can be made by the Children's Hearing System (CHS) or via the Criminal Procedures (Scotland) Act 2005 for remand or sentencing (Malloch 2013). It is current policy that for young people appearing at court, all alternative options to secure care and custody are explored (Lightowler et al 2014). Young people

Table 7.1 Young people's pathways into secure care in Scotland (SASR 2013, 7[3])

Circumstances leading to secure care
The child has previously absconded and is likely to abscond again
If the child were to abscond, it is likely that the child's physical, mental or moral welfare would be at risk
The child is likely to engage in self-harming conduct; or that the child is likely to cause injury to another person

are placed in secure accommodation either because they are deemed as a risk to themselves, to others, or both (for more information on the experiences of young people in secure care in Scotland see Foreman 2004, Foreman and McAllister 2006, Barry and Moodie 2008). Once in secure accommodation, in general no distinction is made in terms of location or treatment of at risk/risky young people, leading Harris and Timms (1993, p.5) to describe it as "precisely the locus in which the question of whether a child has committed an offence or is in some other way problematic, ceases to matter".

Across different local authorities in Scotland, it has been found that use of secure accommodation differs, reflecting not only local circumstances but also local decision making and practice (Walker et al 2006). The Scottish Government is responsible for funding the placements of those young people who are sentenced into secure care for offences, but for placements made on welfare grounds, the financial responsibility lies with the relevant local authorities. The average cost of a secure care bed per week during 2014–2015 was £5,495. In 2014–2015 there were 90 beds available in five secure accommodations services across Scotland. The geographical location of these five units can result in placements distant from home; for example, one study showed that in a research sample of 76 young people, just 21 were placed in a secure unit in their own local authority (Barry and Moodie 2008).

The vast majority of young people in secure accommodation have complex needs and have experienced multiple disadvantages, including: physical and mental health problems; poor educational achievement, school exclusion and truancy; abuse (physical, emotional, and sexual) and neglect; parental mental health / substance misuse problems; experiences of loss and grief; and families with significant criminal histories (Kendrick et al 2008a, Whitehead et al 2010, Mitchell et al 2012). Scottish Government Children's Social Work Statistics for secure care show that as of 31 July 2015, 70 per cent of the young people in secure care were aged 15 or over; 69 per cent were male; and 93 per cent had a known additional support need. A majority of 84 per cent of stays were 6 months or under (53 per cent for up to 3 months). Examining secure authorisations made by the CHS found marked gender differences in decision making, which for girls/young women tended to focus upon behaviours that presented a risk to themselves 'particularly due to sexually risky behaviour' (Whitehead et al 2010, p.4), and boys/ young men for presenting risks to others as well as themselves (Muncie 2004).

Young people who are accommodated in secure settings often have long histories of involvement in the CHS (Kendrick et al 2008b). Research by the Scottish Executive (2009) showed that within a sample of 53 young people in secure accommodation, 83 per cent were admitted from residential units or residential schools. Most had experienced more than one care placement in the year proceeding secure accommodation. In their research, Whitehead et al (2010) found that 74 per cent of their sample had previously had secure authorisations, and 24 per cent had been on the Child Protection Register at some point. This shows that secure care is embedded in the Scottish social work practice and wider forms of care for young people. Through the young people's pathways there are links to different forms of care, but also to other closed establishments like young offender's institutions or psychiatric units within Scotland as well as the wider UK (Schliehe 2014).

Researching secure care

The data acquired for this chapter originates from two doctoral research projects on young women in the Scottish criminal justice system. While both projects have a mixed method approach with a particular focus on in-depth qualitative interviews, one is based in human geography while the other has a criminological outlook. A combination of primary interview data and secondary sources like media reports, press releases and policy documents aids the analysis of secure care environments in Scotland as it presents as a relatively new academic field in both human geography and criminology. With regards to the researchers' positionality, the interview process differed and rather than assuming a 'fixed' position, it had to be re-negotiated in every participant encounter. Research in closed institutions presents its own challenges regarding the researcher's positionality but also issues like access or ethics such as ensuring participant anonymity of a 'captive' young audience (Kendrick et al 2008a). The semi-structured interviews were conducted in three separate empirical phases and the material described in this chapter refers to 11 young women in secure care at the time of interview, 12 young women referring to their experiences of secure care retrospectively and 20 interviews with secure care staff and other professionals working in the field (Schliehe conducted research in 2012 and 2013–2014; Crowley conducted research in 2015). The young interviewees all ranged in age between 14 and 21 years and had experienced varying lengths of time in closed conditions.

Researching secure care environments involves particular dedication to issues of security and awareness of the participants' position and therefore required active listening skills and flexibility to fit the young women's needs and provide them with opportunities to express their views and opinions (Barry 2006, Gaskell 2008). It is beneficial to consult other accounts on interviewing marginalised groups for methodological advice (Curtis et al 2003). Methodologies need to be chosen and adapted with security restrictions and young people's particular needs in mind. The questions to young women ranged from their biographies to how they experience their time in locked environments. Many interviews were complimented by the more creative method of mental map production to be able to understand the young women's pathways (Schliehe 2012). Interviews with secure care staff and other professionals throw up their own challenges, not least negotiating and maintaining a different interview style in the same research environment. In the professional interviews the focus lay on extracting expert opinions on the workings of secure care, the balancing act of providing care under tight security measures, pathways to and through the secure care system, and wider views on the criminal justice system for young people in Scotland.

In presenting the acquired data, this chapter focusses on displaying descriptive accounts to be able to voice young women's and professionals' own accounts and allow their experiences to be foregrounded. In connection with the interview, researchers' own experiences of working in closed environments and their own accounts of participant and spatial observation and 'spending time' in these special research places are important sources. While the empirical data provides new views of life inside secure care and opportunities to find out about otherwise covert environments, this study can only provide a snapshot of the highly complex world of secure care pathways.

Carefully controlled: locking up young people

Being detained in secure care has a profound effect on young people and their lives subsequently. With its highly secure environment and intensive care practice, secure care itself needs to be understood as a closed institution sitting at one end of the scale of a care continuum for most of those detained. It is crucial to further look into life on the inside (see Schliehe 2015) but also to consider young people's pathways beyond secure care (Barry 2006). This is in line with the self-prescribed aim of the unit and its provision of continuing care and transition for young people in order to help rebuild their place within their families and communities that goes beyond the time they spent in the closed institution itself. The following three sections explore (1) internal spaces and structures of secure care, (2) professional views on 'how' and 'why' young people are sent there and their perceptions of the impact of secure care practice and (3) young women's own experiences of moving through (and beyond) secure spaces.

Secure spaces

We are not accustomed to thinking of places of care as unknown, secret places, although secure care units certainly are areas of restricted access. The purpose-built setup and the internal constitution through rules and regulations reveal many aspects of the difficult balance of care and control. After admission a young person's life takes place in this one space. Young people enter into the secure unit through a double-gated vehicle entrance before going into the main building, and receiving an introduction, initial assessment and a search (AC/SC field diary 2014). They sleep, eat and attend school in the same locked building and closed courtyard. Life is governed by routines and a strict regime in order to be able to keep the population of young people safe from themselves and others. Cutlery counts, safe 'care' checks, policies of social interaction and locked doors as well as general vigilance on the part of staff all contribute to managing young people in this closed environment (AS/SC field diary 2014[1]). As described by Mona (AC/SC, residential worker),

> you can't even talk to somebody that you've got a wee friendship going with, you can't even have a private conversation with. There's a staff member going "Excuse me, what are you talking about?"

Every team, according to Sarah (AS/SC, residential staff) works in a multi-disciplinary way to tailor their approach and monitor the young people's progress.

Most staff (AS/SC) mention the structure of secure care as an important characteristic of everyday practice which is accomplished by enforcing routines and regimes within the closed environment. Another of the main aims mentioned is the building of positive relationships to achieve a nurturing environment that helps young people to rehabilitate and re-engage with their own community (AS/SC field diary 2014). Staff explain that young people benefit from strict rules and routines and that it is rather inconsistency that causes young people to 'act out' or 'kick off', with structure considered "really important in this kind of setting" (AS/SC, Julia, residential staff). The rigidity and boundaries in terms of moving around in physical space as well as navigating the social environment create an atmosphere of certainty where "they come in and they know 'right, this is what's happening. I can't just lie on the couch'" (AS/SC, Rachel, residential staff). The rules and regulations governing young people's behaviour and conduct extend to constant and all-encompassing structures of their everyday lives. This concerns what they do, how they do it and where they do it and who they do it with, any hour of the day.

Security is implemented through in-built architectural features as well as in the management of the young people. Many young people and staff describe their particular units as highly secure, not in an overpowering, very obvious way but rather in an efficient, silent and streamlined and automated fashion (AS/SC, Linda, residential staff). There are no high fences around the facilities that would mark them clearly to the surrounding area. The buildings themselves, however, often evoke images of modern 'manicured' fortresses in which attention is diverted to the inside, resisting the outside gaze through purposeful design (AS/SC field diary 2014). Built-in security features are managed from a central control hub. Doors are locked throughout with few exceptions and windows cannot be opened. The courtyards are completely surrounded by the building making it seemingly impossible to climb or escape over the roof – without having to introduce any 'hard' security features like barbed wire or fences (AS/SC field diary 2014). Cameras are installed throughout the building in communal areas and corridors so that incidents can be played back and analysed if necessary.

The unit can broadly be separated into living spaces and 'other' resource spaces used for education, recreation and supplies. The living spaces are organised around the common room with doors going off to corridors and other rooms. The appearance of the living space is different to what might be expected as the common room large windows let light in and the ceiling is high to create a different sense of space (AS/SC, Rupert, residential staff). Walls are often colourfully painted with murals and adorned with photographs of past unit activities such as sports days / celebrations (AC/SC field diary 2014). The common room is created with homeliness in mind but security is never out of sight due to the presence of features of security such as cameras, and the TV locked away in a cupboard:

> the furniture's secure, it's heavier, you can't just pick it up and throw it around. The very fact that there's posters on the walls, there's plants, there's lamps, these wee furnishings take away from how rigid the environment can be and makes it a wee bit more homely' (AS/SC, Rachel, residential staff)

Other rooms and corridors go off from the living area: the locked staff office in each unit is located centrally with a window overlooking the living area. The kitchen is usually locked apart from mealtimes – inside all cupboards are also individually locked. Individual bedrooms are ligature-free spaces with a built-in bed and sideboard, a window that cannot be opened and a bathroom without a door, and opposite the bed is a panel in the wall from which staff can peer in and control the blinds, power and water supply. The bathroom setup is similarly bare, with a showerhead attached to the ceiling, a built-in toilet without toilet seat and a sink without appliances that could be ripped off or damaged easily. The door is big and solid and has no handle on the inside like the window (AS/SC field diary 2014). Contact with the outside world is also carefully controlled. Telephone calls to families can take place every day but are monitored. Visits are also allowed but often distances from home prohibit or at least can make this a challenge for families.

While this brief description of secure spaces and routines in no way extends to all aspects of life and work in secure care, it shows how the setup of these units mirror the complex and conflicting task of enhancing care and management of 'risk' and 'crises' (see also Kendrick et al 2008b, Roesch-Marsh 2014). The task of controlling young people's behaviour in order to care for them is seen as crucial, and is, as such, directly supported by the setup and secure design of the unit.

Entangled pathways – staff perspectives

Staff in secure care settings include social care workers, psychologists, therapists and teachers, who, as part of a young person's individual plan, work together to provide a tailored package of 'interventions' and strategies for their time in secure care. These individual care plans are developed following an initial comprehensive needs assessment on arrival, which includes various screenings and wellbeing checks, and are followed and amended throughout the young person's stay. Approaches such as cognitive behavioural therapy and psycho-education are incorporated to address issues such as problem solving, assertiveness skills, anger management, substance misuse and offending. Depending upon the staff role, work may be specialised – "specific intervention programmes, manualised programmes (…) trauma-informed, forensic-informed" (AC/SC, Georgina, wellbeing staff) – or more general – "my work's the daily stuff, so it's enforcing the routine, making them feel nurtured and kind of addressing their specific issues on a daily basis, like you would do as a parent" (AC/SC, Claire, residential worker).

Staff recognise that secure care often has a 'revolving door' (AC/SC, Emma, wellbeing staff) for young people, who may inhabit this space on more than one occasion. In order to bridge the divide between secure care and community, a number of other 'outside' professionals, including social workers, CAMHS (Child and Adolescent Mental Health Services), voluntary sector workers and health workers also work to provide services both before, during and after a young person's stay. Whilst increasing potential support, some secure care staff interviewed acknowledged that young people may struggle to negotiate and manage the sheer number of relationships with professionals in which they find themselves, and their sometimes competing or overlapping demands and priorities:

> Over time I've learnt to make sure I check, have you got any other workers, and what's going on? Cos things can just start and they'll go to something, and you'd be like, oh it's not overlapping, because you can be saturated you know, because seeing 5 professionals a week, it's a lot. (AC/SC, Mary, wellbeing staff)

In line with other research (Batchelor and Burman 2004, Miller and Mullins 2006, Belknap and Holsinger 2008, Davidson and Chesney-Lind 2009, Daly 2010, Sharpe 2012) the professionals interviewed described how, in their experiences, the pathways of young women into secure accommodation were typically characterised by multiple factors of disadvantage, highlighting in particular, the influence of adverse familial events. There was an acknowledgement that young women in secure accommodation would often (though not always) have been on the radar of statutory services for several years, if not their whole lives, moving through different levels of care with their increasing levels of control. This awareness that professionals have of the backgrounds of young women feeds directly into the approaches taken. For example, poor and/or disrupted attachments were identified as very significant in pathways into secure care, and so staff work hard to build and model positive and healthy attachments. Some staff voiced frustration that these relationships and attachments are then severed upon transitions out of secure care. They talked of how 'unnatural' this is, when compared to an ordinary relationship of continuing reliance: "people don't flit in and out of your life normally" (AC/SC, Cath, residential manager). This same vexation was mentioned in terms of the time restrictions enforced by relatively short stays in secure care, and the limitations that this places on the scope of work carried out, although this was sometimes in conflict with a desire for young people not to be in secure care in the first place (AC/SC, Georgina, wellbeing staff).

For some professionals in decision-making positions, both in and outside of secure care, frustration was coupled with anxiety regarding how they could possibly judge when a young woman would no longer be 'a risk', given the artificiality of secure environments, the "false reality in here" (AC/SC, Chrissy, wellbeing staff). They also acknowledged that by locking a young woman away from access to the resources she was using to commit 'risky behaviours', she would resort to other means, such as self-harming, which in turn would then keep her locked up. Here there seemed to be a conflict between trying to protect a young woman from harm, and needing to address the cause of the behaviour, not its symptoms. These paradoxes of care and control show how secure care environments are contested terrain.

Staff voiced that whilst in secure care, they were able to offer a high intensity care to young people. However, it was recognised that this in turn creates reasons for some to try and stay longer or return (AC/SC, Jennifer, residential worker). They described how some young women will 'act up' as their case review time approaches. There was a common perception that often young people leaving secure care were not supported adequately by social services once back in the community, either immediately post transition, or longer term. Staff expressed dissatisfaction at this (perceived) lack of support, a factor that was seen as placing young people at risk of further custodial or institutional involvement. This was attributed to the heavy caseloads of social workers, or to poor relationships between young people and their individual workers due to factors such as infrequent contact, or an association with negative experiences related to social services (AC/SC, Emma, wellbeing staff). While voluntary sector organisations try to compensate for this, several professionals mentioned how service provision for young people leaving secure care is very patchy with availability depending upon factors like location (rural/city divide) and gender (AC/Community Service, Heather, project manager). It was feared that in an environment of austerity, the availability and quality of such services would decrease further.

> We can build up resilience, we can build up protective factors, we can do all that, but then when the controls of the environment and the staffing groups and the support mechanisms that surround the young person are not there, how do we measure the impact of the work that we've done … when the young person is faced with those issues or these dilemmas when they're out and about? (AC/SC, Abi, residential worker)

Secure care staff and professionals working in the field underline the complicated and entangled pathways that lead young people into secure care. The complex web of measures of care and control employed to cater for and secure the young detainees highlights some of the deep seated contradictions embedded in the notion of caring for adolescents in a closed and restricted environment. As a result, dependencies can be created, paradoxical relationships experienced, and much needed consistencies not maintained.

Moving stories

For many young women, being locked up is part of a longer history of place changes. Their institutional journeys are intersected by many other types of accommodation (e.g. foster care, care homes) which is a main determinant for multiple home changes. Many describe long histories of the care system requiring them to move within the same town or city as well as between different parts of Scotland and Britain. All interviewees in secure care have experienced previous care settings (see Table 7.2) and between one and four stays in secure care – some with added experiences of closed psychiatric units. The young women interviewed in the community (in connection with the voluntary sector organisation 'Up-2-Us' and their 'Time for Change' project that deals with girls and young women at risk of detention [for more information see Burman and Imlah 2012]) have had similar care involvement and mostly multiple experiences of secure care, sometimes leading on to time in prison.

Table 7.2 Overview of placements and moves of detainees in secure care (AS/SC)

Name	*Care placements*	*Secure care stays*	*Psychiatric facilities*	*(Minimum) moves[2] overall*
Ava	2	1	0	5
Gina	2	2	1	7
Maggie	8	1	2	16
Mona	1	1	0	3
Nora	17	2	1	22
Pam	2	1	0	3
Phoebe	2	1	0	3
Rhiannon	14	1	0	15
Suzie	2	1	2	6
Mia	2	4	0	7

Talking of 'countless' care arrangements is a common way of summing up their experiences. Nonetheless distinct trends and movements can be detected in the analysis of these young women's pathways. Looking at emerging patterns in their mobility, it becomes apparent that many seem to reappear in the same or different institutions along their journey. From an age of 16 and higher many of them have experienced stays in homeless accommodation. The experience of 'homelessness' is not tied to age, however, and many young women under the age of 16 have experienced periods of 'rooflessness', 'sofa-surfing' and the like. Most young women also report staying with different family members for longer periods of time. The appreciation of what this highly mobile, insecure lifestyle entails is often missing from analyses of institutional life.

This high mobility often leads to a different attitude towards belongings and attachment to things as well as people. That is the case for young people who have to move far, for example, from England to a Scottish secure unit as well as moving many times within the same city (14 different children's homes all in the Glasgow area; AS/SC, Rhiannon; also AS/TfC, Ruth). Most young women describe an entangled web of heteronomous as well as self-determined orchestration of location changes. Changing 'home' environments means a constant challenge to adapt to new environments, new rules and new people living in sometimes close proximity. The unpredictability of social and geographical surroundings leads to different responses from craving stability and a 'normal' life (AS/SC, Mona) to feeling detached from a concept of 'home' (AS/TfC, Ruth). 'Home' as a geographic location as well as a place for personal possessions and intimate social ties to family members is in many cases a disrupted and undermined concept for these young women. Secure care units oftentimes play a major part in the young detainees' experiences of displacement. Due to the often longer stays in secure care — reportedly months to sometimes years — they work as social and geographical fixtures.

The closed environments with their strict regime of control and high levels of care in and by itself evoke strong emotional responses. Other studies (see Kendrick et al 2008b) support the findings that most young people find their detention in locked environments distressing and upsetting. There is a lot of uncertainty when these young women are asked about their feelings on secure care, acknowledging that it benefits their safety but on the other hand struggling with the restricted contact to the outside world and a lack of autonomy and responsibility (Schliehe 2015). When it comes to a transition period and leaving secure care, other issues emerge. Many young women tell about uncertain time frames and worries about re-integration. Thinking about secure care retrospectively, some young women wonder about getting 'too much care' and getting used to a level of support that makes life in the community without constant staff supervision very challenging:

> I just don't think it's a good place for people to go. I'm no' saying it's bad or anything but in there it's a false environment I 'hink. So you're in there and you're cared for and stuff like that and then when you're out, like, you just want to be back. And I know most of the people who have been in here who have been in the secure done the same, they went back and then they've come to prison. Most of the people outside that I know, they're the same. And it's the attachment you've got to that place. (AS/TfC, Flo)

While the spatial fixture of secure care can hold positives, it can also have detrimental effects on transition due to its disparity from life outside. The changing environments – particularly the strict regulations in closed spaces – create a different sense of dependency and security that can lead to challenges in the community. While some are so used to permanent surveillance that outside spaces increase their perceived vulnerability, others report on increases in public scrutiny and stigma in the community, like by local police forces or people they know (AS/TfC, Lara). High levels of mobility and being locked up/held secure both seem to carry similar stigma making it difficult for young people to move on. Their own sense of mobility and rootlessness is a factor that needs to be taken into account when assessing young people's movements through different forms of care and in particular secure care with its particular disparities of care and control.

Conclusion

The secure unit as an institution is the most structured and controlled environment that young people can be detained in. All of their life takes place in this one place: they sleep, eat, go to school, spend free time in the same locked building and closed courtyard. Life is governed by routines and a strict regime in order to be able to keep the population of young people safe from harm(ing). To be able to provide the right proportion of security and care is a balancing act that is managed with staffing levels and purpose-built spaces. Both security and care features depend on an effective daily schedule. These routines provide a ‘safety net’ of predictability and structure when dealing with ‘un-manageable’ and ‘impulsive’ young people. This control is partly exercised through building relationships with young people – a measure that seems uncertain and fluid compared to the ‘built-in’ features of control, but no less important in the management of the young detainees. In this secure setting, care is part of the overall practices of control both in the practical everyday work within the units as well as in management and planning.

What clearly appears from the empirical data above is the fact that the high ‘care’ mobility is in many cases involuntary and heteronomous in nature. The underlying apparatus of movement often starts a lot earlier in these young people’s lives than stays in closed institutions. There is a noticeable link between mobility and exclusionary societal practices of which the closed institution sits on the far end of the continuum of exclusion. The young people’s own perception of their level of self-determination (or the loss thereof) is closely linked to feelings of rootlessness. In many cases, stays in closed institutions amount to months and sometimes even years at a time. Reflecting on time spent in closed institutions is, however, often only possible in retrospect because detainees cannot be sure how much time they will be held for. Similar to the young people, secure care staff and professionals highlight the entangled pathways leading into secure care and the complex and contradictory nature of providing high-intensity care in highly restrictive and controlled settings.

Looking at the transition from childhood to adulthood, Valentine (2003) refers to the ‘liminal period of youth’ as a time of complex and fluid transitions. The reflexivity of this period can be seen in the different lifestyles, sub-cultures and identities that young people adopt, but Valentine (2003, p.40) recognises a simultaneous increase of risk “in the form of guilt or blame if they end up on the margins of society” as a result of their behaviour. Dealing with ‘dangerous’ youth has taken specific form in secure care with an amalgamation of welfare and punishment that has an enormous effect on the individual detainees. While the larger group of young adults finds itself constructed through negative imagery, this particularly applies to ‘institutionalized’ youth from secure care. Young women in this setting may be viewed with added gravity because of their gender. While wider debates on policy and practice raise contradictory issues around welfare and punishment, these go hand in hand in secure environments combining high surveillance, behavioural control and comprehensive therapeutic care. Rather than being deployed as alternative strategies, secure care combines penal welfare in its intertwining measures of care and control. With this distinct amalgamation, secure care units raise the need for further research into concepts like care itself. Often presenting as what Valentine (2003, p.44) calls ‘care careers’ they find their ‘height’ in secure care units. Looking beyond the immediate institutional environment throws up further questions about dislocation, marginalisation and continuing careers to more ‘grown-up’ institutions.

There is great scope for more research into young people and their experiences of secure care and the geographies of closed spaces in Scotland (see Kendrick et al 2008b, McKellar and Kendrick 2013, Roesch-Marsh 2014, Schliehe 2015), the UK (see Harris and Timms 1993, Bullock et al 1998, Goldson 2002) and similar settings in other countries (Halsey 2006, Andershed 2013, Chesney-Lind and Shelden 2014). Researching secure care environments involves particular dedication to issues of security and awareness of the participants' position and therefore requires the researcher to consider the particular ethical and environmental issues when entering closed settings. Capturing pathways to understand institutional journeys is vital in order to learn more about the interface between institution and young people. The complex nexus of the nature of closed institutions and how they are experienced by detainees cannot be conceived without directing attention to individual pathways. The contraries of care and control are acutely felt by the detainees beyond their time of detention often leading to increases in institutional pull-factors and working against the proclaimed aim of prevention and rehabilitation.

Note

1 Throughout this chapter, AS and AC refer to observations and findings recorded by the authors. SC refers to observations made in Secure Care units, while TfS refers to data from the 'Time for Change' programme which works with young people in closed institutions.

2 It must be noted at this point, however, that 'move' itself is a slightly slippery term and is in this case defined as a move to a different geographical area (like from parental home to care home) that is intended to last a considerable amount of time and is intended as a 'home'. Moves that take up significant time in the young women's lives but are very temporary – albeit sometimes continuous in nature – like stays in police cells or moving while roofless are not included. The way the figures are deducted from the young women's data here, no moves are attributed a length of stay – albeit it needs to be acknowledged that length of stay does differ significantly in individual cases. The length of stay time frame is removed from the data display on purpose for keeping individual anonymity.

References

Andershed, A.K. (ed.) (2013) *Girls at risk: Swedish longitudinal research on adjustment*. New York, Springer.

Barry, M. (2006) *Youth offending in transition: the search for social recognition*. London, Routledge.

Barry, M. and Moodie, K. (2008) *'This is not the road I want to go down': young people's perceptions and experiences of secure care*. Glasgow, Who Cares? Scotland.

Batchelor, S. and Burman, M. (2004) Working with girls and young women. In McIvor, G. (ed.) *Women who offend*. London, Jessica Kingsley, pp.266–87

Belknap, J. and Holsinger, K. (2008) An overview of delinquent girls: how theory and practice have failed and the need for innovative changes. In Zaplin, R.T. (ed.) *Female offenders: critical perspectives and effective interventions*. Burlington, Jones and Bartlett Publishers, pp.3–42.

Bond-Taylor, S. and Jones, K. (2009) Crime, criminalization and criminology: policing trouble? In Bochel, H.M., Bochel, C., Page, R. and Sykes, R. (eds.) *Social policy: themes, issues and debates*. London, Longman, pp.302–28.

Brown, S. (2005) *Understanding youth and crime: listening to youth?* Maidenhead, Open University Press.
Bullock, R., Little, M., and Spencer, M. (1998) *Secure treatment outcomes: the care careers of very difficult adolescents*. Aldershot, Ashgate.
Burman, M. and Batchelor, S. (2009) Between two stools? Responding to young women who offend. *Youth Justice* 9, pp.270–85.
Burman, M. and Imlah, N. (2012) *Time for change: an evaluation of an intensive support service*. Glasgow, Scottish Centre for Crime and Justice Research.
Chesney-Lind, M. and Shelden, R.G. (2014) *Girls, delinquency and juvenile justice*. Chichester, Wiley and Sons.
Collins, D.C.A. and Kearns, R.A. (2001) Under curfew and under siege? Legal geographies of young people. *Geoforum* 32, pp.389–403.
Curtis, K., Roberts, H., Copperman, J., Downie, A. and Liabo, K. (2003) 'How come I don't get asked no questions?' Researching 'hard to reach' children and teenagers. *Child and Family Social Work* 9, pp.167–75.
Daly, K. (2010) Feminist perspectives in criminology: A review with Generation Y in mind. In McLaughlin, E. and Newburn, T. (eds.) *The SAGE handbook of criminological theory*. London, Sage, pp.225–46.
Davidson, J. and Chesney-Lind, M. (2009) Discounting women: context matters in risk and need assessment. *Critical Criminology* 17, pp.221–46.
Dorrer, N., McIntosh, I., Punch, S. and Emond, R. (2010) Children and food practices in residential care: ambivalence in the 'institutional' home. *Children's Geographies* 8, pp.247–59.
Foreman, L. (2004) *'In-secure': the views and experiences of young people in secure accommodation in Scotland*. Glasgow, Who Cares? Scotland.
Foreman, L. and McAllister, S. (2006) *'I wish I could turn back time'*. Glasgow, Who Cares? Scotland.
Gaskell, C. (2008) 'Isolation and distress'? (Re)thinking the place of emotions in youth research. *Children's Geographies* 6, pp.169–81.
Goldson, B. (2002) *Vulnerable on the inside: children in secure and penal settings*. London, Children's Society.
Halsey, M.J. (2006) Negotiating conditional release. *Punishment and Society* 8, pp.147–81.
Harris, R. and Timms, N. (1993) *Secure accommodation in child care: between hospital and prison or thereabouts?* London, Routledge.
Kendrick, A., Steckley, L. and Lerpiniere, J. (2008a) Ethical issues, research and vulnerability: gaining the views of children and young people in residential care. *Children's Geographies* 6, pp.79–93.
Kendrick, A., Walker, M., Barclay, A., Hunter, L., Malloch, M., Hill, M. and McIvor, G. (2008b) The outcomes of secure care in Scotland. *Scottish Journal of Residential Child Care* 7, pp.1–13.
Lightowler, C., Orr, D., and Vaswani, N. (2014) Youth justice in Scotland: Fixed in the past or fit for the future? Glasgow, Centre for Youth and Criminal Justice.
Malloch, M. (2013) *In their own right: support for families with a young person in secure accommodation*. Edinburgh, Families Outside.
McKellar, A. and Kendrick, A. (2013) Key working and the quality of relationships in secure accommodation. *Scottish Journal of Residential Child Care* 12, pp.47–57.
Miller, J. and Mullins, C.M. (2006) The status of feminist theories in criminology. In Cullen, F.T., Wright, J.P. and Blevins, K.R. (eds.) *Taking stock: the status of criminological theory*. New Brunswick, Transaction, pp.217–50.
Mitchell, F., Roesch-Marsh, A. and Robb, L. (2012) *Taking stock of alternatives to secure accommodation or custody for girls and young women in Scotland*. Glasgow, Criminal Justice Social Work Services.
Muncie, J. (2004) *Youth and crime*. London, Sage.

Roesch-Marsh, A. (2014) 'Out of control': making sense of the behaviour of young people referred to secure accommodation. *British Journal of Social Work* 44, pp.197–213.
Rose, J. (2002) *Working with young people in secure accommodation: from chaos to culture*. London, Routledge.
Scottish Executive (2009) *Securing our future: a way forward for Scotland's secure care estate*. Edinburgh, Scottish Executive.
Scottish Government (1995) *The Kilbrandon Report. Children and Young Persons, Scotland*. Edinburgh, Scottish Government.
Scottish Government (2016) *Children's Social Work Statistics Scotland 2014–2015*. Edinburgh, Scottish Government.
Schliehe, A.K. (2012) *'It rips my knittin' – girls and young women in the Scottish criminal justices system. Beyond total institutions?* Unpublished MRes dissertation, University of Glasgow.
Schliehe, A.K. (2014) Inside 'the carceral': girls and young women in the Scottish criminal justice system. *Scottish Geographical Journal* 130, p.71–85.
Schliehe, A.K. (2015). Locking up children and young people – secure care in Scotland. In Evans, B., Horton J. and Skelton, T. (eds.) *Play and recreation, health and wellbeing: geographies of children and young people*, vol. 9, New York, Springer.
Secure Accommodation (Scotland) Regulations (SASR) (2013). Edinburgh, Scottish Government.
Sharpe, G. (2012) *Offending girls: young women and youth justice*. London, Routledge.
Valentine, G. (1995) Angels and devils: moral landscapes of childhood. *Environment and Panning D* 14, pp. 581–99.
Valentine, G. (2003) Boundary crossings: transitions from childhood to adulthood. *Children's Geographies* 1, pp.37–52.
Walker, M., Barclay, A., Malloch, M., McIvor, G., Kendrick, A., Hunter, L. and Hill, M. (2006) *Secure accommodation in Scotland: its role and relationship with 'alternative' services*. Edinburgh, Scottish Executive Education Department.
Weller, S. (2006) Situating (young) teenagers in geographies of children and youth. *Children's Geographies* 4, pp.97–108.
Whitehead, I., Henderson, G., McNiven, G. and Lamb, D. (2010) *Secure authorisations in Scotland's Children's Hearings System: an exploration of decision making, placements, outcomes and social backgrounds*. Glasgow, Scottish Children's Reporter Administration.

8 Children creating spaces of care in diverse early childhood centre built environments

A complex interplay of social relations and materialities

Ann Pairman and Carmen Dalli

Introduction

Developments in the conceptualisation of spatiality mean that space is no longer viewed as a pre-existing 'container' that determines people's social lives. Instead space is understood as a fluid product of interactions between social relations and materiality (Massey 1992), a notion captured in Soja's (1989, p.79) term 'spatiality'. In this chapter we use Lefebvre's (1991) notion of the spatial triad and Gibson's theory of affordances (1979) to demonstrate how 'spaces of care' were produced in four early childhood (EC) centres, as children and teachers interacted with, and within, each built environment. We argue that diverse built environments (the material) intertwined with people's conceptions of space to influence children's mobility, social interaction, and agency.

Kraftl (2006) has noted that since space is continually socially constructed, the spaces of children's institutions bear the evidence of cultural values, practices and power relations beyond their physical dimensions, including particular understandings or constructions of 'childhood'. EC centres are such institutions; whether purpose-built, or located in premises modified from their original function – such as family homes, church and community halls, or commercial warehouses – their physical dimensions, or built environment, bear evidence of the interplay of pedagogical and cultural values with spatial practices. However, there has been a relative lack of attention paid to materiality in childhood research. This may reflect the dominance of social constructionism within the sociology of childhood. Alan Prout (2005) has argued that accounts of the 'socially constructed child' privilege a discourse that tends to be silent or vague about materiality, "whether it be architectures, artifacts, technologies, or other" (2005, p.63). However, in recent years, children's geographies has imbued childhood studies with a "sense of spatiality" as part of a shift toward materialist perspectives (Holloway and Pimlott-Wilson 2011, p.13).

Researchers examining the lives of children in EC centres have begun exploring children's perceptions of space use (Einarsdottir 2005b), bodily, sensory, and affective experiences of space (Horton and Kraftl 2011), interactions between pedagogical values and children's spatial experiences (e.g. Rutanen, 2012, Vuorisaloa et al 2015), and power dynamics inherent within such dimensions (e.g. Gallacher 2005). However, it is difficult to find literature that specifically examines links between the configuration and size of built environments and children's experiences. This chapter aims to contribute to the shift towards materialist perspectives by drawing on a research study examining relationships between aspects of the built environments in four New Zealand early childhood centres and 2- to 5- year-old children's lived experiences, casting light on children's production of 'spaces of care' in diverse built environments.

Geographies of care

The notion of care is central to early childhood pedagogy. With a history rooted in the need of mothers working outside the home for custodial care of their children, traditionally the focus has been on how children in group-based early childhood services are physically cared for. More recently, Conradson (2003) points to the opening up of the geographies of care to encompass broader definitions of care and diverse care environments. Within children's geographies, this has included a much greater consideration of children themselves as carers (Evans 2012, Robson 2004) and the mutually constituted experiences of care as a result of child and adult interactions (Gallacher 2005). These shifts have been supported more broadly by a strong line of literature on a feminist ethic of care that has impacted on childhood sociology and education (e.g. Goldstein 1998, Noddings 1984). These studies have opened up much more complex notions of care and show how care is woven into the fabric of social spaces and communities (Conradson 2003).

A key focus in many studies of care in EC centres has been on how teachers show care to children. In this regard, Goldstein (1998) explains that caring "is not something you are, but rather something you engage in, something you do" (p.247) acting in the best interest of the one cared-for. Among other things, this requires empathetic and respectful engagement in responsive and trusting relationships. Much less common is a focus on spaces of care within EC settings and how physical design and space impacts on children's sense of care and well-being. Spaces that afford children the opportunity to act *in care of themselves* (for example by enabling privacy when children seek it) project an image of trust and respect between adults and children and position children as beings with agency rather than entirely dependent on adults (Clark and Moss 2001, James et al 1998). This chapter interprets 'spaces of care' as being spaces children can produce themselves to respond to their own social, physical, or emotional rhythms, that is, spaces where they can act in 'in care of themselves'.

The context of this study is New Zealand [NZ] EC centres. NZ has had an integrated care and education policy structure since 1986 (Dalli 1990, Meade and Podmore 2002) when administration and policy responsibility for child-care services was transferred from the Department of Social Welfare to the Department of Education. The move established the important principle that care and education are inseparable and led the way to the introduction, in 1988, of 3-year integrated training in colleges of education for childcare and kindergarten teacher trainees. Having a shared training background for staff working across all types of EC services for children aged 0 to 5 years has broken down traditional barriers between the childcare and kindergarten/preschool sectors and brought about greater awareness that learning results from all experiences including ones that may be seen as primarily related to children's physical care. This is reflected in NZ's curriculum framework, *Te Whāriki,* which defines curriculum as "the sum total of the experiences, activities, and events, whether direct or indirect, which occur within an environment designed to foster children's learning and development" (Ministry of Education 1996, p.10).

Research on EC built environments

Traditional research on children's spatial experiences in EC centres arose from the field of environment and behaviour sciences (Moore 2004). As noted earlier, much of this research is underpinned by a view of space as a 'container', although several studies have highlighted the interplay between the social and the material (e.g. Moore 1986). This largely quantitative body of research provides useful insights into the ways the arrangement of furnishings may help shape children's experiences. For example EC physical environments with clearly defined activity areas are associated with more exploratory behaviour, positive social interactions, cooperation among children (Moore 1986), and increased independence and competence (Maxwell 2007, Trancik and Evans 1995). However, Nichols-Whitehead and Plumert's (2001) finding that children in highly organised physical environments were less likely to transform spaces also points towards the significance of the construction of spaces for power relations between adults and children. As Kritchevsky and Prescott (1969) argued in an early paper on this topic, the availability of 'potential space' also contributes to children's ability to act on space.

The ability for children to 'act on space' can also be linked to the literature on child-friendly space. Since the 1980s the debate about how space shapes behaviour has moved towards questions about what child-friendly spaces might look like, and how they can be created. Definitions of child friendliness typically recognise the significance of children having agency to respond to their own social and physical rhythms (e.g., Broberg et al 2013, Horelli, 2007). Chatterjee (2005) drew on literature about friendship to define a child-friendly place as one that, among other factors, "promotes exploration and actualization of its many affordances for different activities and social interactions…allows children to express themselves freely in creation and control of territories and special places" (p.17).

Drawing on Roger Hart's (1979) seminal work on children's experience of place, David Kennedy (1991) describes child-friendly spaces as ones where children can act with agency and "leave their mark on the world" (p.44). Children's friendships and peer interactions are integral to these understandings of child-friendly spaces and literature on children's relationships highlights the importance of interplay between children's agency, mobility, and social interaction. Corsaro's (1988) work on peer culture drew attention to children's movement (spatial practices) and materials. He found that preschool children attempted to protect ongoing play from intrusion by manipulating the material environment and claiming ownership over space. Similarly, Einarsdottir (2005b) noted that Icelandic children enjoyed occupying rooms they could move into and "close the door and play without adult interference" (p.470).

Children's ability to control their levels of social interaction is an important aspect of care. As Manning-Morton (2014) pointed out: "if we consider the amount of time we expect very young children to be part of large (often noisy) groups, including opportunities for being alone should surely be part of planning practice and provision" (p.21). Literature on crowding suggests that the configuration of the built environment plays a significant role in such provision. Theoretical literature on crowding links two concepts to the feeling of 'being crowded': excess of stimulation and lack of privacy (Gove et al 1979). Altman's (1975) privacy regulation theory views privacy as '"boundary regulation"', a process whereby individuals or groups control others' access to them. The configuration of the built environment may moderate the effects of crowding by influencing patterns of space use. For example, Evans et al (1996) found that people felt less crowded in complex spaces in residential homes. Complex spaces were described as having more areas to pass through to get from one room to another.

There is little research related to young children's feelings of being crowded in EC centres although research has examined children's strategies when withdrawing from social contact. For example, Skånfors et al's (2009) ethnographic study of 2- to 5- year-old children's withdrawal behaviours in a Swedish EC centre found that children wishing to withdraw from social contact made themselves 'inaccessible' by "acting distant, reading books and hiding" (p.100). These strategies appeared to be connected to the centre's daily rhythms, meaning that children withdrew at busier times, possibly to avoid feeling crowded.

The recent and growing trend in EC research to focus on very young children's spatial experiences (e.g. Einarsdottir 2005a, Gallacher 2005, Horton and Kraftl 2011, Raittila 2012, Rutanen 2012) possibly reflects the influence of children's geography on childhood studies which increasingly view children as "social actors who construct their own childhood spaces" (Mannion 2007, p.409). Some of this work picks up on power relations in young childen's spatial practices. For example, Rutanen's (2012) study of the experiences of 1- to 3-year-old children in a Finnish EC centre found that adults constrained children's use of space during activities with pre-defined objectives and during transitional times with adults using children's age to justify these differentiated spatio-temporal practices. However, during activities with flexible objectives and locations, Rutanen found that space was open to moments of negotiation between adults and toddlers (2012).

This concept of negotiated space is present also in Gallacher's (2005) study of forty-five 2- to 3-year-old children in a Scottish nursery which found that toddlers "appropriate and reconfigure space and time for themselves", suggesting a dynamic power relationship. Gallacher concluded that the "official world organised by adults and the peer culture 'underlife' [were] intricately intertwined" and constantly jostled together (pp.258–9). This indicates that very young children are not only able to establish a peer culture but also to exert agency in relation to spatial practices.

In childhood studies, the term *agency* is widely used with its meaning often assumed rather than directly defined (Guo and Dalli, in press). Barker (2008) notes that agency is broadly understood as a person's capacity to act independently and make 'free' choices. While childhood sociology views children as 'endowed' with agency, it is also acknowledged that the skills and competencies which enable children to act with autonomy and agency develop over time (Ghirotto and Mazzoni 2013) and are also influenced by the context and its structures such as class and gender (Lee et al 2013). In this way 'images of childhood' act as a structure that influences children's agency. For example, Vuorisalo et al (2015) illustrated how adults' images of childhood and pedagogical approaches (such as differentiations between 'free play' and teacher-led activity) played a role in enabling or constraining children's participation and the creation of relational spaces in Finnish EC centres.

While this chapter considers how adults' images of childhood influence the production of social space, the key focus is on how the size and configuration of built environments intertwine with those images, and how together they influence children's agency, mobility, and social interactions as they produce spaces where they are in care of themselves.

Lefebvre's spatial triad

Henri Lefebvre's (1991) spatial triad offers a particularly useful theoretical lens through which to examine how space is produced within EC built environments. Lefebvre's triad unites three epistemologies about space and sees these as interwoven and working together simultaneously in relation to any event or social practice. They can be compared to "three cameras projecting simultaneously" (Zhang 2006, p.222). The three 'camera' lenses are: (1) material spatial practice or *perceived space*; (2) representations of space or *conceived* space; (3) and representational space or *lived* space (Lefebvre 1991). This last lens, lived space (the space as it is directly lived) includes everyday appropriations of place that also bring into being alternative imaginations of space (see discussion of affordances in the next section).

In relation to EC centres, *perceived* space includes things that can be perceived by our senses (e.g. seen, smelt, or touched). This includes people's actual movements, furniture arrangements, locations, openings, visibility and so on. *Conceived* space includes evidence of socially constructed meanings, such as signs, cultural codes, or rules that may represent particular understandings of 'childhood' or 'learning'. An example may be the way furnishings and spaces are arranged to indicate their expected use. Although adults hold the most power in the construction of conceived space, children contribute when they create their own rules and expectations as part of their peer culture. *Lived* space illuminates the space "as it is directly lived through its associated images and symbols" (Lefebvre 1991, p.38). It is the space of children's lived experiences, and includes meaning making, feelings, and understandings of the space as people encounter it. In an early childhood centre, it would include children's actual use of space and their feelings about the window that's too high, the toilet with no door, or the corner that enables intimacy. Used as research tools, each of Lefebvre's 'camera lenses' illuminates a different perspective on space; when the lenses are used together, without any one lens privileged over another, it becomes possible to understand how social space is produced. The notion of social space is similar to Soja's (1996) concept of 'thirdspace', in that it gathers the perceived physical environment, the cultural codes, meaning and rules, and the lived space together.

Gibson's affordances

The interaction between the three dimensions of Lefebvre's conceptual triad is not linear but intertwined. To help understand this interplay, the study drew on Gibson's theory of affordances (1979). An affordance is the *opportunity* an aspect of the physical or material environment offers to an organism. Affordances exist in the perceived space and may be obscured or made visible by the conceived space. For example, a bookcase affords a child the opportunity to climb or jump off (provided the child is physically capable). Climbing on or jumping off are *potential* affordances or possible actions. An affordance may be perceived (or seen) but not acted on by the child in deference to the rules, expectations or cultural norms of a particular context, for example rules about 'what a bookcase is for' (conceived space). Alternatively the affordance may not be perceived because the child has no interest in climbing or because cultural norms obscure its existence. Similarly the act of utilising an affordance (e.g. climbing the bookcase) can influence the conceived space. For example, a teacher's response to children climbing a bookcase contributes to the ongoing creation of rules and expectations that feed into the production of social space.

The study

The data we draw on in this chapter were collected by the first author and come from an ethnographic, mixed-method, multiple-case study which investigated 2- to 5-year-old children's experiences in four spatially diverse early childhood centres. The data-gathering tools used are consistent with the key principle behind the 'mosaic approach' of doing research with children (Clark and Moss 2001). In other words, they resembled ordinary practices in EC settings and included: child observations via video and voice-recorded notes that were subsequently transcribed; spatial mapping of children's locations and activities; child-led tours; the use of stories, photos, and photo books to engage children in conversation; teacher focus groups; and general discussions with teachers and children.

In each centre, data were gathered over a period of 2 to 4 weeks and included: continuous observations of 3 to 6 focus children for one full day each, with the rest of the children in the centre also observed as they participated in activities or conversations in response to their interests and the researcher's focus. All but two of the children attending the centres were part of the study (total average attendance: 125 children across four centres). Of the focus children, at least one each was two years old, three years old, and four years old.

The spatial mapping combined time sampling and photographs from set vantage points to pinpoint people's locations every 20 minutes for two days in each centre. The photo books incorporated the image of a well-known children's book character into photos of different spaces within each centre. Children were asked to tell the character how various spaces were used, identify special places and spaces they used when they wanted to be alone or quiet, and tell the character about centre rules and so on. On some days, data were collected over the entire opening hours of the EC centre (10 to 10.5 hours per day), for example 'whole-day' observations of individual children and photos that contributed to spatial mapping. These data were analysed holistically within and across the four case studies to identify key themes (Berg 2007) in keeping with the research questions which guided this research.

The case study EC centres

The four case study EC centres (Table 8.1) were situated in NZ's capital city, Wellington, and were selected on the basis of each built environment's potential to contribute to an 'information rich' group of centres. Each EC centre operated all-day (at least 50 hours per week) and under-2-year-olds could not self-access the space used by the 2- to 5-year-olds (the study group). Two of the centres had larger indoor activity spaces (3.25 m^2 per child), and two were smaller (NZ's minimum standard of 2.5 m^2 per child). One each of the large and small centres had an 'open' and visually simple spatial configuration (all activity spaces could be seen from one vantage point), or a 'complex' and enclosed spatial configuration (activity spaces could not all be seen from one vantage point).

More than 80 per cent of adults in each centre were qualified EC teachers holding either a 3-year diploma or degree, and all programmes were based on child-led activity and free-flow play (Bruce 2001) during which children moved freely between spaces (including the outdoor) for long periods of time. In this chapter we have drawn together data excerpts from the four case study centres and put them together as a multi-perspectival bricolage (Denzin and Lincoln 1999) which is interpreted reflexively to draw out threads of meaning, or themes. Although the case study design aimed to gain in-depth, 'rich descriptions' situated within each case, some comparisons are made between cases in the discussion that follows.

Table 8.1 Spatial features of case study built environments

Centre SC Small, complex[1] configuration. Situated on the ground floor of a renovated domestic home. Comprises three activity rooms. Two rooms are situated at either end of a larger 'main' room (one enclosed, one semi-enclosed). Average attendance: 23 2- to 5-year-olds.	**Centre SO** Small, open[2] and visually simple configuration. Situated in a 'purpose built' building with a series of rectangular 'classrooms'. Areas in the study comprise two classrooms. Average attendance: Class 1: 31 2.5- to 3.5-year-olds. Class 2: 23 3.5- to 5-year-olds.
Centre LC Large, complex configuration. Situated in a single story renovated domestic home. Comprises five activity rooms, two with wide openings between. Two enclosed rooms (called *Fern* and *Flax*) are situated at either end of a large L shaped hallway. Average attendance: 24 2- to 5-year-olds.	**Centre LO** Large, open and visually simple configuration. Situated on the ground floor of a renovated industrial building. Comprises one large T-shaped activity area. Part of the outdoor is covered and accessible during rain. Average attendance: 24 2- to 5-year-olds.

[1] Activity spaces could not be seen from one vantage point.
[2] Activity spaces could be seen from one vantage point.

Spaces of care: materiality and social relations in diverse built environments

In this section we use three contextualised cameos created from data excerpts to illustrate differences in the children's experiences of their built environments. The differences point to an interplay between the materiality of the built environments, the children's use of this materiality, and the teachers' approach to children's agency within the constraints of the material space.

The first cameo is set in centre LC (large, complex) and begins in the early afternoon on a partly rainy day. Leo (4-year-old boy) had engaged in sedentary play with his friend Jimmy for most of the morning. During a dispute about turn-taking, Leo managed his emotional response by physically withdrawing from all social interaction and moving into a secluded space for a short period. In the afternoon Leo became increasingly physical and moved outside where he appeared to disagree with the other children's rules for a game and returned indoors, moving directly into the Fern room (small, enclosed room).

Through a Lefebvrian lens, the Fern room's *perceived space* includes walls that separate it visually and auditorily from other spaces. The mirror in this room reflected Leo's image and the absence of people created a quiet space devoid of other movement. The quality of the perceived space afforded Leo opportunity to withdraw socially to engage in physical and imaginary play uninterrupted, and for Leo and Hugo to occupy this space privately together. Leo's experience of the Fern room (*lived*

Box 8.1 An intimate space of care

Figure 8.1 'The baby's trying to attack me'

Observational field notes record: "*Leo is completely alone and spinning his body around and around while holding a doll in front of him. He's looking in the mirror and says 'the baby's trying to attack me'*" (see Figure 8.1).

Leo appears lost in his imaginary world. Leo tells Ann, the researcher, that he's come into the room because he has cold hands but his choice of an unoccupied space, among other warm rooms available, also suggests to Ann that he wants to be alone. Teachers do not enter the room during this time as Angie (teacher) explained: "*it's almost as if there's an agreement that those two rooms* [Fern and Flax] *are kind of – not adult free – but that they're kind of owned by the kids more than the adults*". After 15 minutes of playing alone Leo's friend Hugo arrives at the entrance but does not enter. Leo looks at Hugo and smiles saying "*you've got the same shirt as me, I've got a shirt like that at home*". Hugo enters the space and they begin talking to imaginary creatures in the mirror together. Leo says "*look he's doing the same as me… I need the army of Sharman because I'm Sharman*". (Sharman appeared to be an imaginary character, possibly from popular media, but was unknown to the researcher). Hugo calls the 'army' through the mirror (see Figure 8.2) saying "*come on, we need you*". The boys continue this imaginary game and have the room to themselves for half an hour before five children arrive from outside.

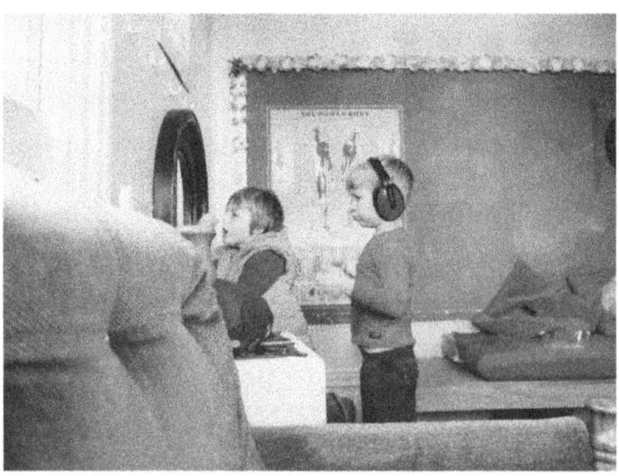

Figure 8.2 Calling the army of Sharman through the mirror

space) was one of quiet withdrawal from social interaction, followed by intimate social interaction that involved co-constructed fantasy with an 'invited' friend.

However, Leo's lived space did not arise directly in response to the affordances inherent in the perceived space. As Lefebvre argues, three aspects of the spatial triad work together and, although not explicit in the cameo, the *conceived space* (the children and adult's conceptions of this space) influenced Leo's ability to perceive and utilise the room's affordances. Rules and cultural codes governed children's spatial movements as they responded to the emotional, social, and physical aspects of their daily lives. At LC, teachers adopted a knowing but flexible management style of the children's movements allowing them agency in their spatial practices. Angie (teacher) explained,

> they own the space and they can use it in a way that they want to.... Children have got the choice and the freedom to use space independently and they know how to do that because they've been provided with the culture and the support to ... have privacy.

These unwritten rules appeared to be understood by the children and were reflected in the way they conceived of the Fern room as being available as a private and quiet space. Joe, another 4-year-old boy in the same centre as Leo, said he liked the Fern room because it was "dark and quiet" and he was observed occupying it alone on several occasions, sometimes further enclosing himself by climbing inside a 'postbox' that he had placed on the floor. Similarly, Tom, also 4 years old, reflected that "when I come in here it's quiet" while Rose and Lily (4-year-old girls) explained that they often went in there with just one or two other friends. Clearly, the conceived and perceived spaces combined to enable Leo to produce a quiet and private 'lived' space of care where he responded to his own emotional and physical rhythms by acting 'in care of himself'– an experience shared by other 4-year-olds in his centre.

These children's experiences of creating 'spaces of care' at LC were similar to those at centre SC (small, complex) which was the only other centre with walls that separated some spaces visually and auditorily from other spaces. In the smaller but still complex centre children were frequently observed in enclosed spaces alone or with one friend. However their experiences contrasted with those of the children in centre SO (small, open).

In centre SO, the perceived space was visually and auditorily open. There were no full height barriers so a sense of privacy was only partially afforded by the arrangement of furniture and noise travelled easily. Furthermore, centre SO's teachers felt that parents and management expected children to be actively supervised at all times. These values, or conceptions of what was appropriate behaviour in relation to children's safety, were reflected in the design of the centre's environment which ensured that children could be seen from one vantage point.

At centre SO older children appeared to have difficulty finding indoor spaces that provided them with quiet or privacy when needed. For example, when asked where they went if they wanted to be alone or in a quiet space with one other friend, most children identified outdoor spaces. Amy (a 4-year-old girl) explained "because all the spaces have children in it" with her friend Wai adding "[we] don't have any quiet spaces". Liam was observed saying to a teacher "I need somewhere quiet to go by myself" before crouching against a wall beside a 'play oven'. Video data showed that the room remained noisy and the space did not afford Liam privacy. In this very open, small built environment children's experiences (lived space) appeared to be imbued with a sense of congestion and noise, and a lack of privacy or intimacy.

The ability of children to occupy space alone or in pairs for prolonged times points to the importance of centre size as well as complexity of the space for providing children the opportunity to create spaces of care for themselves. Because there was a lot of potential space (Kritchevsky and Prescott 1969) children at centre LC could usually find space to occupy in order to control their levels of social interaction or protect interactive play (Corsaro 1988). In addition to this, LC's complexity and size simultaneously enabled groups of children to follow their own rhythms in relation to physically boisterous or quiet play. This was especially evident on a rainy day and is demonstrated in the second cameo, which picks up where the first cameo left off.

Box 8.2 A boisterous space of care

After spending half an hour alone in the room Leo and Hugo seem positive about the other children's entry and together they transform the Fern room into a site of boisterous, loud group play. The game 'Army of Sharman' evolves into one where all seven children have superpowers. They organise themselves into goodies and baddies and begin jumping around the room and climbing on the furniture and windowsills. In field notes from a similar event at LC the researcher, Ann, noted that "*Although they're shouting and being fairly loud it is not particularly audible from other rooms – apart from the hallway. I was not aware of this play until I actually came into the hallway area.*"

After a few minutes all seven children leave the room together – superpowers intact. They move through the hallway and into the adjoining rooms, congregating where a teacher is reading a story to the children. They jump around on the three couches in this area and continue to act out their play. Some throw themselves around on the carpet and make 'phew phew' noises. The teacher says "*we're trying to read here perhaps you could take your superpowers outside*" (see Figure 8.3). Jessie (4-year-old girl) says "*no we're not going to play outside because we're playing inside*" (they have recently returned from playing outside in the cold). The teacher replies "*no, it's making it hard for us to read*". One of the boys says "*we're going outside*" but Jessie replies, "*no, no we [are] going in that room*", and the group of seven run back to the Fern room (see Figure 8.4) and continue their boisterous play. As their play no longer disturbs the story the teacher accepts the children's decision to stay inside.

Figure 8.3 Superpowers disturb the story

Figure 8.4 Back to the Fern room

As noted earlier, in both LC and SC, which had complex built environments, there is a degree of auditory separation between rooms. However in contrast to the experiences at LC described in cameo 2, the small size of centre SC meant that noise still travelled between rooms because of their proximity to each other. Karey (teacher) explained "if it gets … loud it accelerates and accelerates … it feels like the whole space then has that ripple effect through". In response to these elements of the perceived space, teachers at centre SC governed the children's mobility indoors in the interests of the group as a whole. For example, the enclosed room at SC had rules about its use and teachers usually asked children to leave the space, or to act more quietly, when they became noisy or boisterous. Rules and codes comprising the conceived space influenced children's experience of this room (lived space) with one child observing that "you've got to be good in there". By comparison, the children's boisterous and loud activity at LC had very little impact on other indoor spaces because the overall built environment was large with the hallway space between rooms, and the existence of more rooms, minimising the noise effect. This allowed teachers at LC to choose not to prevent loud play in enclosed rooms, thus affording children the opportunity to occupy different indoor spaces in response to their own rhythms in relation to levels of social interaction and physicality.

The teachers at SC expressed some ambivalence about how much to limit boisterous play inside, but they all talked about the need to mitigate noise and manage the group's 'rhythms' through the day. Teachers recognised that access to boisterous, quiet, social, and solitary experiences was an important feature of children's well-being and care and, because the small space sometimes constrained children's ability to create spaces of care that enabled them to follow their own rhythms, teachers responded by gently governing the rhythms of the group through a combination of 'hui' (regular children's meetings), 'resets' (reorganising the physical space), and creating meal-time rituals such as lighting candles. These actions provided powerful codes about spatial practices (conceived space) producing social spaces (lived space) that reflected a less dynamic power relationship between adults and children than occurred at centre LC. As Karey (teacher) explained, "it is a long day, like it's a really long day, 5.30pm for some children, so it really has to have peaks and troughs".

The situation in centre SO's small, open space differed markedly from that in the centres with complex built environments. SO's simple rectangular configuration meant that noise in one part of the room had a significant effect on all other parts of the room. Children and teachers expressed concerns about high noise levels, and video recordings clearly demonstrated the way noise 'travelled'. LO's open, but larger 'T' shaped configuration sometimes reduced the impact that noise in one space had on another but, particularly on wet days, concerns about noise influenced teachers in both spaces to take a stronger role in governing children's mobility and noise levels to support the well-being of the *group as a whole*.

While LC's large, complex built environments, and to a lesser degree SC's smaller environment enabled many older children to 'act in care of themselves' by controlling their own rhythms through their spatial practices, complex built environments had quite different implications for younger children's ability to create spaces of care with peers.

Two-year-old children have long been considered highly mobile compared to older children. Despite Arnold Gesell describing toddlers' mobility in a 1975 study, research on this topic has remained scarce (Johansson et al 2015). Younger children's constant movement was evident in the data from this study and raises interesting questions about how younger children's mobility interacts with the built environment to afford them control over their social interactions and the ability to act 'in care of themselves'. Two-year-old children appeared to gain more control over their peer relationships and mobility in open built environments than in environments with complex space and multiple rooms.

This final cameo provides an excerpt from a 2-year-old child's everyday life at centre LO (large, open). Compared to peer relationships among 2-year-olds in centre LC, peer relationships among 2-year-olds in centre LO were extremely strong with very frequent examples of younger children 'travelling together', usually with buggies, handbags and other items (we revisit this point in the next section). This is significant because when younger children lead or 'frame' the play, they can be seen as acting with agency and participating on their own terms (Emilson and Folkesson 2006). The following cameo takes place on a sunny day and begins at 3.10pm. Ella has been at centre LO since 8.30am.

This spatial practice of 'travelling' or 'journeying' was observed in 2-year-olds in all four case study centres, even in the absence of 'buggies'. However, at LO and SO 2-year-old children commonly *journeyed with peers.* In contrast, in the two centres with complex layouts this spatial practice seemed to be constrained by the visual limitations of the space. For example, 2-year-old children at LC (large, complex) tended to move from room to room in quick succession. In one video segment two-year-old Angelo moved into seven visually separate spaces in a 3-minute period. The 2-year-olds at LC tended to make these 'journeys' alone. Significantly, when these children did attempt to 'journey together' video episodes and observation transcripts show them appearing to 'lose' each other. Similarly, at centre SC 2-year-old children most often interacted with peers and led their own play when outside, where they appeared better able to maintain visual contact.

Box 8.3 Journeying to a space of care

Ella (2 years old) makes two journeys, using a buggy, to transport a toaster, jug, and other items from the play kitchen to the locker area at the other end of the large T-shaped activity space. She spreads out the equipment on the floor saying "*I'm going to have breakfast here*". Ella walks towards the other end of the space, from where she can see the entire space, calling out to her friend "Adam" several times. She finds Adam outside and says "*come with me our Adam*". They 'journey' together back to the locker area and Ella sits by the toaster and says to Adam "*come and join me*". Adam sits down. After watching a parent collect their child from the nearby infant room, Ella observes that "*baby Louis has picked up his mum*", then says to Adam "*I've got some toast, do you want some toast*?" Ella watches Billy's mum arrive to collect Billy and says "*mummys daddys picking me up*".

Esme (also 2 years old) sees Ella and Adam from the far end of the centre and walks over to join them. After a short dispute over the equipment, Ella asks Esme "*do you want to come out for dinner*?" Esme, Adam, and Ella play 'dinners' for a few more minutes, before knocking on the infant room's glass door. A teacher lets them in for a visit. This appears to be a regular routine for Ella and Esme who were observed asking to enter this room towards the end of the day on several occasions.

From a Lefebvrian perspective, centre LO's perceived space was large and open with substantial amounts of empty (or potential) floor space. This afforded children visibility across spaces and opportunity to easily walk (or run) in pairs or small groups without distraction. In addition, the locker area was conceived by teachers as being 'out of bounds', resulting in the space being unoccupied and void of equipment prior to Ella's arrival. The resulting potential space allowed Ella to occupy and transform it while affording her a clear line of sight to the entry door, the under-2-year-old room and, importantly, towards other indoor spaces and her friends (see Figures 8.5 and 8.6).

Figure 8.5 View from picnic area to under-two-year-old room

Figure 8.6 View from picnic area to entrance and main play area

This interplay between the conceived (out of bounds) and perceived space enabled Ella to create a space of care with her peers. In this space these highly mobile toddlers were able to utilise the visual connection afforded by the open space to journey together, or join others as they congregated in this area. Teachers at centre LO were aware of the toddlers' movement patterns. Tess (teacher) compared the LO's open areas to the more complex renovated house she had previously worked in. She said "if you're in closed off spaces I don't think you have that free space which a lot of our young ones [have]… I think this space here allows [2-year-olds] to run straight through" … [Alice (teacher) adds] "they get those trolleys and they go 90 miles an hour". Possibly in appreciation of toddlers' desire to 'journey' teachers sometimes turned a 'blind eye' to their occupation of the 'out of bounds' locker area.

These data suggest that visual connection between toddlers is important in maintaining these younger children's peer interactions. Other studies have previously described visual connection with adults as influencing toddlers' movements and social experiences. For example, Legendre and Fountaine (1991) reported that toddlers spent more time engaged in social play with peers in semi-enclosed spaces when there were visual barriers between themselves and the adults. The study we have discussed in this chapter adds the insight that visual connection *among toddlers* also has an influence on their peer interactions: they help toddlers to maintain their 'journeying together'. Furthermore, the toddlers' tendency to constantly move through space may afford toddlers greater opportunities for social interactions relative to more complex environments.

Conclusion

The focus of this chapter has been on exploring the idea that diverse early childhood built environments afford different opportunities for children to respond to their own social, physical, or emotional rhythms and act 'in care of themselves'. The data we have called on comes from an investigation by the first author of 2- to 5-year-old children's experiences in four diverse early childhood centre built environments. To unpack the nature of the children's experiences, we created a bricolage of data excerpts from the four case-study centres and presented them as contextualised cameos. We reflected on these cameos using Lefevre's (1991) theoretical construct of space as a triad, together with Gibson's (1979) notion of affordances.

The analysis in this chapter clearly shows that children's agency to create 'spaces of care' was strongly impacted by a complex interplay between the affordances of the materiality of the built environment (the perceived space) and its socially constructed meanings about what was allowed or seen as possible by the teachers and the children (the conceived space). In particular, the analysis points to differences in the ways that toddlers and older children utilised the affordances of the built environments to produce spaces of care with their peers. Where older children made full use of the affordances of complex built environments to create intimate and private spaces of care together, 2-year-old children benefited from the affordances of open spaces where they could visually connect with their peers and engage in 'journeying together'. It was also clear that, overall, the centres of a larger size, and including more potential space, created more opportunities for children to: experience privacy; follow their rhythms; be boisterous; and generally be in care of themselves without infringing on the care other children might be seeking for themselves.

This analysis about children's experience of space goes beyond the view of space as merely a 'container'; by using the Lefebvrian notion of space as a triad, it contributes to the growing literature on the notion of spatiality as well as to understandings of geographies of care, specifically related to children being in care of themselves in early childhood centres. It points to a need for further research that explores how built environments influence young children's lived experiences, and in particular children's ability to act in care of themselves. Our chapter has drawn on data from four NZ case studies and provides detailed analysis of children's spatial experiences in those cases. The insights gained from this study provide a beginning for further study to develop a broader base of knowledge capable of informing policies and practices related to geographies of care. For example, the data reported in this chapter suggests that NZ's regulated minimum requirement for EC centre indoor space (2.5m^2 for children 0 to 5 years) constrains the physical affordances available to children, such as auditory separation and potential space, which are important to children's ability to act in care of themselves. However a wider research base is needed to justify policy shifts at a government level.

Children in countries other than NZ may share similar experiences of EC built environments to those in this NZ study. In 2011, the OECD's average minimum standard for indoor activity space was 2.9m^2 for over three-year-olds and 3.6m^2 for zero to three-year-olds (Pairman 2012). The adequacy of the US standard (3.2m^2 or 35 ft.) has long been of concern (White and Stoecklin 2003). A NZ trend whereby minimum standards appear to be less likely to be exceeded due to property costs and the privatisation of the EC sector (Bedford and Sutherland 2008) may also be mirrored internationally. Such trends indicate an increasingly urgent need for further research.

However, as demonstrated in this chapter, the interwoven nature of Lefebvre's spatial triad, and the interplay between elements of the triad and Gibson's notion of affordances, remind us that the regulation of perceived space (size or configuration) can only create the *conditions for high quality* experiences for children in EC centres – a point made by Goelman et al (2006) in relation to the other so called structural elements of quality, such as adult:child ratios. To understand the interplay between social relations and materiality in EC settings, further research is needed that focuses on how children's everyday lives are played out in EC centres, as spaces for 'living'. Such research could usefully investigate how children's spatial practices are influenced by each EC community's values about children's lived childhood (such as supervision), and by the rules and norms that arise from these in different contexts. Additionally, there is clearly scope for this research to be carried out by researchers from across the fields of children's geographies, childhood studies, and environment and behaviour.

Acknowledgment

The research reported in this chapter was supported by a Victoria University of Wellington PhD Scholarship awarded to Ann Pairman.

References

Altman, I. (1975) *The environment and social behavior*. Monterey, Brooks/Cole.

Barker, C. (2008) *Cultural studies: theory and practice*. London, Sage.

Bedford, M. and Sutherland, K. (2008) Early childhood education and care in New Zealand: do we have a runaway train? *The First Years: Nga Tau Tuatahi. New Zealand Journal of Infant and Toddler Education* 10, pp.38–42.

Berg, B. L. (2007) *Qualitative research methods for the social sciences*. Boston, Pearson.

Broberg, A., Kyttä, M. and Fagerholm, N. (2013) Child-friendly urban structures: Bullerby revisited. *Journal of Environmental Psychology* 35, pp.110–20.

Bruce, T. (2001) *Learning through play: babies, toddlers and the foundation years*. London, Hodder and Stoughton.

Chatterjee, S. (2005) Children's friendship with place: a conceptual inquiry. *Children, Youth and Environments* 15, pp.1–26.

Clark, A. and Moss, P. (2001) *Listening to young children. The mosaic approach*. London, National Children's Bureau.

Conradson, D. (2003) Geographies of care: spaces, practices, experiences. *Social and Cultural Geography* 4, pp.451–54.

Corsaro, W. (1988) Peer culture in the preschool. *Theory into Practice* 27, pp.19–24.

Dalli, C. (1990) Early childhood education in New Zealand: current issues and policy developments. *Early Child Development and Care* 64, pp.61–70.

Denzin, N.K. and Lincoln, Y.S. (eds.) (1999) *The SAGE handbook of qualitative research*. London, Sage.

Einarsdottir, J. (2005a) Playschool in pictures: children's photographs as a research method. *Early Child Development and Care* 175, pp.523–41.

Einarsdottir, J. (2005b) We can decide what to play! Children's perception of quality in an Icelandic playschool. *Early Education and Development* 16, pp.469–88.

Emilson, A. and Folkesson, A.-M. (2006) Children's participation and teacher control. *Early Child Development and Care* 176, pp.219–38.

Evans, G.W., Lepore, S.J. and Schroeder, A. (1996) The role of interior design elements in human responses to crowding. *Journal of Personality and Social Psychology* 70, pp.41–6.

Evans, R. (2012) Sibling caringscapes: time–space practices of caring within youth-headed households in Tanzania and Uganda. *Geoforum* 43, pp.824–35.

Gallacher, L. (2005) 'The terrible twos': gaining control in the nursery? *Children's Geographies* 3, pp.243–64.

Gesell, A., Ilg, F.L. and Ames, L.B. (1975) *Infant and child in the culture of today*. London, Gesell Institute of Child Development.

Ghirotto, L. and Mazzoni, V. (2013) Being part, being involved: the adult's role and child participation in an early childhood learning context. *International Journal of Early Years Education* 21, pp.300–8.

Gibson, J. (1979) *The ecological approach to visual perception*. New York, Cornell University.

Goldstein, L. (1998) More than gentle smiles and warm hugs: applying the ethic of care to early childhood education. *Journal of Research in Childhood Education* 12, pp.244–71.

Goelman, H., Forer, B., Kershaw, P., Doherty, G., Lero, D. and LaGrange, A. (2006) Towards a predictive model of quality in Canadian child care centers. *Early Childhood Research Quarterly* 21, pp.280–95.

Gove, W., Hughes, M. and Galle, O. (1979) Overcrowding in the home: an empirical investigation of its possible pathological consequences. *American Sociological Review* 44, pp.59–80.

Guo, K. and Dalli, C. (in press) Belonging as a force of agency. An exploration of immigrant children's everyday life in early childhood settings. *Global Studies of Childhood.*

Hart, R. (1979) *Children's experience of place.* New York, Irvington.

Holloway, S.L. and Pimlott-Wilson, H. (2011) Geographies of children, youth and families: defining achievements, debating the agenda. In Holt, L. (ed.) *Geographies of children, youth and families.* London, Routledge, pp.9–24.

Horelli, L. (2007) Constructing a theoretical framework for environmental child-friendliness. *Children, Youth and Environments* 17, pp.267–92.

Horton, J. and Kraftl, P. (2011) Tears and laughter at a Sure Start Centre: preschool geographies, policy contexts. In Holt, L. (ed.) *Geographies of children, youth and families.* London, Routledge, pp.235–49.

James, A., Jenks, C. and Prout, A. (1998) *Theorizing childhood.* New York, Columbia University Press.

Johansson, E., Hagströmer, M., Svensson, V., Ek, A., Forssén, M., Nero, H. and Marcus, C. (2015) Objectively measured physical activity in two-year-old children – levels, patterns and correlates. *International Journal of Behavioral Nutrition and Physical Activity* 12, pp.2–7.

Kennedy, D. (1991) The young child's experience of space and child care center design: a practical meditation. *Children's Environments Quarterly* 8, pp.37–48.

Kraftl, P. (2006) Building an idea: the material construction of an ideal childhood. *Transactions of the Institute of British Geographers* 31, pp.488–504.

Kritchevsky, S. and Prescott, E. (1969) *Planning environments for young children: physical space.* Washington, DC, National Association for the Education of Young Children.

Lee, W., Carr, M., Soutar, B. and Mitchell, L. (2013) *Understanding the Te Whariki Approach. Early years education in practice.* London, Routledge.

Lefebvre, H. (1991) *The production of space.* Oxford, Blackwell.

Legendre, A. and Fontaine, A.-M. (1991) The effects of visual boundaries in two-year-olds' playrooms. *Children's Environments Quarterly* 8, pp.2–16.

Manning-Morton, J. (2014) Thinking about well-being in early childhood. In Manning-Morton, J. (ed.) *Exploring well-being in the early years.* Buckingham, Open University Press, pp.13–27.

Mannion, G. (2007) Going spatial, going relational: why 'listening to children' and children's participation needs reframing. *Discourse: Studies in the Cultural Politics of Education* 28, pp.405–20.

Massey, D. (1992) Politics and space/time. *New Left Review* 196, pp.65–84.

Maxwell, L.E. (2007) Competency in child care settings. The role of the physical environment. *Environment and Behavior* 39, pp.229–45.

Meade, A. and Podmore, V. (2002) *Early childhood education policy co-ordination under the auspices of the Department/Ministry of Education. A case study of New Zealand.* Paris, UNESCO.

Ministry of Education (1996) *Te Whariki: He whariki matauranga mo nga Aotearoa: early childhood curriculum.* Wellington, Learning Media.

Moore, G.T. (1986) Effects of the spatial definition of behavior settings on children's behavior: a quasi-experimental field study. *Journal of Environmental Psychology* 6, pp.205–31.

Moore, G.T. (2004) *Environment, behaviour and society: a brief look at the field and some current EBS research at the University of Sydney.* Paper presented at the 6th International Conference on Environment-Behavior Studies, Tianjin, China.

Nichols-Whitehead, P. and Plumert, J.M. (2001) The influence of boundaries on young children's searching and gathering. *Journal of Cognition and Development* 2, pp.367–88.

Noddings, N. (1984) *Caring. A feminine approach to ethics and moral education*. Berkeley, University of California Press.
Pairman, A. (2012) The relationship between the physical environment and learning: a blind spot in New Zealand early childhood education discourse? *New Zealand Annual Review of Education* 21, pp.21–45
Prout, A. (2005) *The future of childhood*. London, Routledge.
Raittila, R. (2012) With children in their lived place: children's action as research data. *International Journal of Early Years Education* 20, pp.270–79.
Robson, E. (2004) Hidden child workers: young carers in Zimbabwe. *Antipode* 36, pp.227–48.
Rutanen, N. (2012) Socio-spatial practices in a Finnish daycare group for one- to three-year-olds. *Early Years: An International Journal of Research and Development* 32, pp.201–14.
Skånfors, L., Löfdahl, A. and Hägglund, S. (2009) Hidden spaces and places in the preschool: withdrawal strategies in preschool children's peer cultures. *Journal of Early Childhood Research* 7, pp.94–109.
Soja, E. (1989) *Postmodern geographies. The reassertion of apace in critical social theory*. London, Verso.
Soja, E. (1996) *Thirdspace: journeys to Los Angeles and ether real-and-imagined places*. Oxford, Blackwell.
Trancik, A.M. and Evans, G.W. (1995) Spaces fit for children: competency in the design of daycare center environments. *Children's Environments* 12, pp.43–58.
Vuorisaloa, M., Rutanen, N. and Raittilaa, R. (2015) Constructing relational space in early childhood education. *Early Years* 35, pp.67–79.
White, R. and Stoecklin, V. (2003) *The great 35 square foot myth*. Available at www.whitehutchinson.com/children/articles/35footmyth.shtml
Zhang, Z. (2006) What is lived space? *Ephemera* 6, pp.219–23.

9 Children and young people as providers of care

Perceptions of caregivers and young caregiving in Zambia

Caroline Day

Introduction

Research on young caregiving in the global North has drawn attention to the negative (and sometimes positive) outcomes that caring for a parent (or relative) with a disability or chronic illness may have on youth transitions to 'independent adulthood' (Evans 2010, Evans and Becker 2009). The concept of young people caring challenges western constructions of childhood and youth, which believes young people should be protected from adult responsibility (Evans and Becker 2009, James et al 1998). As a result, much of the caregiving undertaken by young people is hidden within the home (Barry 2011, Smyth et al 2011). Certainly research from the UK has indicated that young caregivers may in effect be 'silenced' due to the risk of discrimination or stigmatisation, and also the fear that they will be separated from their families and placed under the registration of child protection and child welfare authorities (Barry 2011, Wardale 2003, Aldridge and Becker 1993). Its hidden nature (Becker 2007) also relates to the fact that many young people do not necessarily identify themselves as young carers due to the nature of the caring relationship and societal norms surrounding caregiving (Smyth et al 2011). This suggests that actual numbers may only represent a proportion of the real number of children and young people caring for someone at home.

A growing literature has revealed the substantial regular care work that young people may undertake for family members affected by HIV and other disabilities in Eastern and Southern Africa (Evans 2012a, 2012b, 2010, Bauman et al, 2006, Evans and Becker 2009, Robson 2004, Skovdal 2011). In Africa, it is widely accepted that children and young people take on responsibility within the household, often at a much younger age than that expected in the global North. In Zambia, children are regularly expected to help at home with cooking, cleaning and chores (especially girls) from as young as five or six years old as part of their 'training' towards being independent and 'good' adults that will make a good husband or wife for someone (personal communication, May 2010). To some degree children in many parts of Africa have long held responsibility for tasks that are not expected of children and young people in the global North. It is often entirely normal that young people work (Ansell 2005), and work includes a variety of activities including street work, domestic work, mining, vending (Bourdillon 2000) and increasingly, caring.

Despite this, as recently as 2009 (p.24), Evans and Becker wrote "there is little recognition of the distinct role of children with caring responsibilities in sub-Saharan Africa, particularly the extent, nature and outcomes of their caring tasks, and there are few services specifically designed to support them". While the body of literature focusing on HIV is now extensive, where young people are concerned it has tended to focus on orphanhood and vulnerability (Foster et al 2005) and the social impacts of HIV (Evans 2010). Combined with the hidden nature of care, social and cultural expectations, as well as the lack of recognition of household care in policy and practice, this has meant there are still no official figures for the number of children and young people involved in caring in sub-Saharan Africa (Becker 2007, Robson et al 2006). While some small country-specific studies have gone some way to estimate the number of young carers present (Robson et al 2006), due to the hidden nature of care, alongside social and cultural expectations, a precise number is far from available. This chapter therefore attempts to expand understandings of how care is often hidden in other contexts beyond the global North.

The hidden nature of caregiving, as evidenced in the global North (Evans and Becker 2009), is a key consideration when examining the occurrence of young caregiving in the global South, particularly Africa, where it could be argued that young people who care are both visible and hidden at the same time. They are visible in the sense that socio-cultural expectations of care, family relationships and gender roles have actively encouraged, indeed expected, and therefore widely recognised that care is undertaken within the family. This means it is quite accepted for a young person to have some caring responsibility and is not seen as anything 'out of the ordinary'. Indeed young people often see their caring role as 'just part of what they do' as part of their family obligation, and therefore not worthy of additional attention. However, it has hidden the number of young people who are increasingly taking on demanding care roles for, or in the absence of, parents or other relatives, from active recognition within society as well as policy and practice. The scale of care required and the lack of government initiatives in many African countries mean that care becomes further hidden within the work of the home (Bowlby et al 1997, Milligan 2005). In addition to this both disability and HIV are heavily stigmatised in many African communities, preventing people from actively seeking support (Evans and Day 2011, Evans and Becker 2009, Ogden et al 2006, Muyinda et al 1997). As a result, care as a 'pressing concern' is often marginalised from view (Lawson 2007) and young people's active roles in providing care for parents or relatives are overlooked (Evans 2010).

Older young people, or youth, are often further hidden and their responsibilities recognised even less (Day 2016, Day and Evans 2015). Few studies on care have extended their research to include 'youth', with the focus instead on *children* and those aged under 18 years. It is therefore important to recognise that many *youth* are also living in households in which "adults are either not present or not fully functioning, in terms of providing material, practical and/or emotional support for children and young people" (Payne 2012, p.294). While an increasing number of studies acknowledge their research to be with children *and* young people, the majority are still based on research with children. While there are a few existing studies of young adult carers over the age of 18 (Evans and Becker 2009), most are small-scale and based on a local population (Dearden and Becker 2000). While the reason for this is not clear, it could be assumed that this is dictated in some way by both policy and practice. In terms of practice, this may be due to their age as well as perceived levels of maturity, knowledge, skill development and the fact they have had longer to develop coping strategies and networks of support, which mean it is assumed they will cope better in situations of adversity. In terms of policy, most funding at government or NGO level is targeted at individuals up to the age of 18 and rarely over this (Potter et al 2012). This fails to recognise that many caregiving youth originate from child- or grandparent-headed households and may have had to care for family members in the past – ‘critical moments’ (Thomson et al 2002) that may still affect their transitions to adulthood. This chapter furthers discussion of care in the global South by drawing attention to the need for a more relational conceptualisation of young people’s life transitions that are characterised by interdependence between family members and the ‘continuum of care’ that shapes young people’s pathways to adulthood, recognising these are often hidden behind wider cultural, social and economic expectations and pressures.

Research context and methods

This chapter discusses the findings of qualitative research that explored the experiences of young people (aged 14–30) who had significant family caring responsibilities and those without such responsibilities in Zambia. The Zambian Government defines ‘youth’ as young people between the ages of 18 and 35 (MSYCD 2006), although the UN definition of youth (15 to 25 years) is also recognised (United Nations 2007). My research revealed that few people identified themselves as ‘youth’ after the age of 30 as they were usually employed, married and had children and hence 30 was used as the upper age limit. The slightly lower age limit reflects the fact that many young people had significant responsibilities at a much younger age than age-based definitions of youth. Semi-structured interviews and life-mapping methods were conducted with a total of 35 young people (aged 14–30 years), comprised of 15 young people who were caring for a chronically sick or disabled parent or relative (9 female, 6 male) and 20 who did not have specific caring responsibilities for a family member (9 female, 11 male), although they often still made significant contributions to the household economy in terms of income generation (particularly the young men) and unpaid reproductive work. The semi-structured interviews with young people were complemented with a life-mapping activity where young people developed a timeline of their lives to map “self-defined ‘fateful moments’ of transition” (Worth 2009, p.1052) from their past and that may occur in the future.

Of the young people with caregiving responsibilities, five cared for their sick or disabled mother, two were caring for their brother, two for their grandmother and two for their uncle, while others cared for their father, grandfather, sister, aunt and niece. The research did not specifically ask for details about the disability or illness of the relative, however when this information was volunteered it was clear the young people were having to care for and manage a range of conditions. Four family members reported HIV-related illness, although the chronic illness of several other family members was also likely to be HIV-related. Other conditions reported included stroke, blindness, spina bifida, cerebral palsy and hydrocephalus, epilepsy and amputation. Those who were not caring for someone sick or disabled were often still responsible for other family members who needed financial support and emotional guidance, such as younger siblings, nieces and nephews. Many young people had witnessed the death of parents or family members earlier in their lives, often from HIV-related illnesses.

Semi-structured interviews were also conducted with 12 parents or family members, eight of whom were being cared for by one of the young people. Interviews were also conducted with 12 key informants from NGOs, support institutions and government departments working across the areas of disability, HIV and children and youth. All the participants' accounts have been anonymised to protect their identities.

Understandings and perceptions of young caregivers

With young caregivers being central to this study it was important to ascertain people's awareness of them and understanding of their responsibilities. Generally, awareness of young caregivers was low amongst all the participants in this study, although there was some acknowledgement that they did exist. When trying to ascertain people's awareness of the existence of young caregivers and who they understood to be caring, there was an overwhelming focus, particularly from the professional and strategic interviews, on 'children' not 'youth'. Despite emphasising the focus on 'youth' in interviews, the majority of participants would consider the question from the perspective of 'children'. Children were considered less able to cope with the responsibility of a caring situation and therefore in need of most attention, compared to youth who were assumed because of their age to have developed a wider skill set that would enable them to cope better. Often, however, young people's roles in providing care were not recognised at all, were considered a rare occurrence or were described as 'not normal'. Professionals still widely believed that 'adult' members of the extended family were the ones who took on caregiving roles, despite recognition, in line with research by Evans (2013) and Evans and Becker (2009), that family structures were changing and many traditional caring relationships were either no longer there, or were coming under increasing pressure in the current economic context.

Few strategic individuals, whether working directly with young people or within organisations that supported the wider family, described having direct contact with caregiving youth, apart from those who had a significant amount of contact with youth who they were aware had significant responsibility at home. One individual, who managed a Youth Resource Centre to provide vocational skills training for young people not in education, said their staff often worked with young people who had stopped attending school because of their responsibilities at home.

Staff at the centre therefore tried to understand each individual's circumstances and make allowances for this:

> Our rules are not that strict on admission because we know some are caring for sick parents, so they may be found at the hospital for long hours and they come late. But what we encourage them is to tell us their story so we know that this is what is happening in their home.

Where young people were caring, professionals recognised they may have to give personal care to parents and relatives, such as washing, feeding, assisting them to use the toilet or giving them medicines, as well as supporting the family through the earning of an income, finding food and clothing and taking care of the home. A government officer from the Ministry of Sport, Youth and Child Development commented, "their care is not simply giving them medication, but it also includes fending for the family, taking care of the family, in terms of buying food and other requirements in the home". This supports findings from previous studies conducted with young caregivers (Evans and Becker 2009).

Young people and their family members, even those caregiving themselves, were often unaware of any other young people living in their communities who also had a caring responsibility. It was only if they had a community care role that there awareness was sometimes heightened. One young woman (age 30) described having contact with other caregivers through her role as a Dorcas Mother in the Seventh Day Adventist Church. This is a group of women in the Seventh Day Adventist Church who offer community service to those vulnerable and in need. Here she had channelled her own experiences of caring for her mother and niece into becoming a mentor for other carers, giving 'advice and guidance' to educate others about how to care for someone else and improve care in the home.

The reasons for the lack of awareness of other caregivers were not explicitly discussed; however, findings from other research studies indicate a number of potential factors. The stigma of illness and disability (particularly HIV) and the associated stigma attached to caring for someone with such a condition may have prevented young people from making their caregiving roles known (Evans and Day 2010, Evans and Becker 2009). Having significant responsibilities at home may have encroached on the free time available for young people to socialise with others and meet young people in similar situations to them and if they did meet they may not have wanted to discuss their situations at home with others (Evans and Becker 2009). It could also be that because it was expected that caregiving would be undertaken within families the young people did not see it as additional to their family obligation and therefore not worthy of additional attention. This reflects understandings that care is founded on love, emotional attachments and reciprocal kinship responsibilities (Becker 2007) as well as reciprocal rights and responsibilities as part of an 'intergenerational contract' to provide care and support during sickness and old age (van Blerk and Ansell 2007, Laird 2005, Collard 2000).

Generally, it was perceived that all young people had responsibilities within their homes, often from a young age, that also increased with age. This supported the notion that young people in many African communities take on an increasing number of household responsibilities as they grow from children to young adults in preparation for one day having their own homes, as well as in support of the family (Evans and Becker 2009, Miller 2005, Katz 1993, 2004, Punch 2001, Solberg 1997). Responsibilities were often based on the assertion of gender and its associated characteristics, which could include, although were not limited to, influences such as physical strength, traditional cultural expectations and perceptions of appropriateness of roles. It did not go unnoticed in the research that a larger proportion of young women in the sample were involved in caregiving within the family than young men, reflecting the gendered nature of caregiving and greater likelihood that girls and young women are called on to provide care, as other studies have found (Evans and Becker 2009, Robson 2004).

Caregiving was understood by professionals to increase the level of responsibility experienced by a young person, both physically and emotionally, often occurring at an earlier age than would normally be expected. The main concern was the financial impact on young people of not being provided for by a parent or relative. The Youth Resource Centre manager commented, "one [challenge] is they lose their source of livelihood. Parents are supposed to provide for them but they are sick or they are dying and they don't provide for them." It was recognised that young people in this position may therefore experience difficulties attending school, or even withdraw from education which posed a concern in terms of the impact on their future opportunities and their ability to find employment:

> These young people, because of the huge responsibility of taking care of their sick parents, grandparents or other relatives, most of them withdraw from school; and then even those that continue going to school find that their progress in school actually is not very impressive because their attention is always divided. (Officer, Ministry of Sport, Youth and Child Development)

Professionals were concerned that young people may leave education in order to find employment to support the family, that young female caregivers could be encouraged into prostitution or having 'sugar daddies' to provide for them, while young men would turn to stealing and crime:

> This forces them to start looking for other means. Where the girls they go into prostitution and having boyfriends who wish to provide for them. For the young men and others they go into stealing, just to survive, that is one of the problems. (Youth Resource Centre Manager)

There was also a fear that young people were taking on complicated caring roles, but without adequate skills or expertise. Few young people were thought to receive training on administering medication, offering personal care or dealing with infections or illnesses which could be infectious. In addition caregiving was recognised as an emotionally demanding role that could cause a young person stress and worry.

> I think a young person caring for a sick [person], it deprives them a lot of their chance to maybe develop well or even just to grow up normal, because you have this huge responsibility of trying to look after the sick. That is why there is an HIV theme that they say "is it how they are infected or affected". If you are not infected maybe you are affected by someone who is... I think it somehow affects you emotionally. (Officer, Ministry of Sport Youth and Child Development)

Caregiving, while perceived therefore as an issue only affecting a small minority of Zambia's youth, was something that was recognised as existing and that could cause challenges for the young people involved.

Young people's perceptions of their caregiving responsibilities

As outlined in the methodology, the young people were caring for a range of relatives, including parents, grandparents, aunts, uncles and siblings with a broad range of ailments. The focus of this study was not to establish the precise illness or disability of the parent or relative, but to understand the impact on the young person and their transitions to adulthood. Most of the young people caring had assumed their roles as caregivers over time. Only two young women had been taken into families to fulfil a caregiving role. The remaining young people had gradually taken on responsibilities as their own caregivers health deteriorated, for example as parents or grandparents aged or became infirm.

The young people perceived their caring responsibilities as 'just what they did' and an intrinsic part of their role within the family. This reflected understandings that caring is part of a reciprocal and loving relationship (Rose and Bruce 1995), as well as part of the 'intergenerational contract' to provide care and support during sickness and old age (van Blerk and Ansell 2007, Laird 2005, Collard 2000). However, in line with research by Milligan and Wiles (2005), none of the young people had 'chosen' to care, but found themselves predisposed to caregiving as a result of the wider 'critical moments' (Thomson et al 2002) that had occurred in their lives, as well as the lack of other practical options, financial constraints (Milligan and Wiles 2005) or simply because there was "no one else to do it" (Bupe, aged 15).

The young people's experiences of care varied. Personal care was not something that had to be provided by all the young people living with a sick or disabled parent or relative. Often the family members being cared for were still relatively active individuals who could still manage certain tasks themselves. Some though were entirely dependent on the young people for everyday tasks such as washing, dressing, getting in and out of bed, going to the toilet and even feeding themselves. For others caring did not always take place as an everyday, permanent activity, but on an 'as needed' basis. This could be because someone's illness fluctuated in severity and they had periods of being well, or because the condition may only arise sporadically, for example, someone suffering from epilepsy. However, while physical care was not necessarily being provided at the time of the research, it did not mean that the burden of care was taken away (Evans and Becker, 2009). The caring young person also had to deal with the emotions and strain of never knowing when the condition might arise, or worsen, and so always being available in case it did. While this may allow the young people periods of respite, the emotional worry and strain never really went away.

Personal care was generally undertaken by females in the household for both men and women. There were instances where girls had to do personal care for male relatives, but in no case did I find a male offering personal care for a female relative. One mother, for example, commented that there was no expectation of her son to care for her when she was ill and that this responsibility fell purely to her daughters. Where daughters were not available it was expected that other family members would come and stay to undertake that role, even if from long distances away. An example of this was Mercy (aged 16) who had inherited the responsibility of caring for her aunt from her sister as a result of her aunt only having sons. A source of care for her had therefore been sought from the wider family – first her sister until she married and now Mercy herself. In several families I was told that brothers or young men did little personal care as it was not appropriate for them to do that for a woman. This supports Evans and Becker's (2009) findings in Tanzania and the UK and recognises the inherently gendered nature of caring as highlighted within Tronto's ethic of care in which women do the caring about, taking care of, and caregiving towards others (Tronto, 1993).

Many of the households were also sites of multiple care relationships, as highlighted in the wider theoretical discourses of caregiving by Tronto (1993), Sevenhuijsen (1998) and Bowlby et al (2010). It was unusual to find a young person caring for one individual, although the other people they were caring for might not necessarily be sick (although they sometimes were). For example, someone looking after a sick parent or relative might also be caring for younger siblings, or the children of deceased siblings. Chisala (aged 30), for example, had remained at home to care for her aging and chronically sick mother. Following the death of her sister she was now also caring for her niece who suffered from epilepsy. She described the daily worry that one or both of her relatives might fall ill. This also had an impact on the level of household responsibilities Chisala had to undertake at home in order to compensate for her relatives now doing less – she had prevented her niece for example from collecting water for fear she might have a fit and fall down the well. In addition to this neither her mother nor niece were well enough to cultivate or to water the fields. She commented, "when you have cultivated you water the garden alone, my mother doesn't water nor my niece water the garden". Her mother and niece made up for this however, by doing more in the house, such as washing dishes and cleaning, which gave Chisala the freedom to focus on other activities, such as cultivating vegetables which she would sometimes take to the Lusaka markets to sell. This did not mean that caring did not have its challenges. Chisala described the responsibility she felt for being the only income generator in the family and not being able to manage like other families. She commented, "caring, one can care, but to manage completely like the way others manage, it becomes difficult. One is not working, that is a problem also. For you to find money for you to use so that you take care of her that's a problem." Joseph (aged 22) was also caring for the children of his deceased brother, as well as caring for his sick mother. He described how he had to contribute towards their education costs as well as providing food and clothes for them. He therefore worked on their own farm to grow vegetables for their own consumption and to sell. He also worked on other people's farms locally to generate additional income.

The caregiving that was described to me was also rooted in notions of reciprocity. Reciprocity occurs between the carer and the cared for, although it may not always be immediate and may involve exchanges over very long time periods and in different forms (Tronto 1993, Sevenhuijsen 1998). What was apparent when talking to the young people was that while they may not be reciprocated immediately for their care, they hoped to be in terms of future rewards, either in terms of inheritance, in the eyes and actions of God, through gaining good fortune, or by gaining experience that would help them in the future to find work. Caregiving therefore, was generally seen as a positive activity. It made the young people feel proud about themselves to know they were doing something good for others. In line with Evans and Becker's (2009) cross-cultural study in Tanzania and the UK, some, however, also described it as a challenging, and a 'burden' that restricted their freedom.

Young people and a 'continuum of care'

One of the significant findings to come from this research was the difficulty that arose when trying to characterise, or label, young people based on their caregiving responsibilities. The research demonstrated the inappropriateness of putting young people into either a caregiving or non-caregiving 'box'. The young people had all experienced multiple events and 'critical moments' (Thomson et al, 2002) over their lives which had changed and shaped the nature of the relationships they experienced with their families and therefore the types of care they both gave and received. Some had cared for family members in the past, but were no longer doing so at the time of the research, while others who were living with grandparents were likely to face this challenge in the future. Caregiving, therefore, had influenced all the young people's lives and experiences at one time or another as they had grown up. While the focus of this research was on those caregiving for a sick or disabled parent or relative, other forms of care, such as looking after siblings, had existed in the majority of young people's lives in line with the ethic of care that expects a person to "strive to meet the demands of caring that present themselves in his or her life" (Tronto 1993, p.126). Many of the young people were also striving to uphold the tradition that care work falls within the domain of the family (Robson et al 2006) when their family networks were often being redefined and reconstituted as a result of losing parents and key family members to HIV and other illnesses. Caregiving was therefore something that all young people had involved themselves in at one time or another, whether through choice or circumstance, and would have potentially impacted on their attitudes towards and experiences of transitions to adulthood. Young people move in and out of a continuum of care over many years which shapes their pathways to adulthood. As a result the distinction of caregiving and non-caregiving youth is not always appropriate in a sub-Saharan African context.

For all young people, caregiving was often the result of other critical moments in their lives, and therefore one of many influences that impacted on and interrupted their transitions to adulthood. All the young people had faced bereavement, family breakdown, poverty and disrupted attendance at school, and it was these factors that often led to young people's involvement in care work. None of the young people had 'chosen' to care (Milligan and Wiles 2005), but found themselves predisposed to caregiving as a result of their situations. Due to the gendered nature of caregiving and women's assumed 'natural' roles as nurturers, young women were more likely to be drawn into care-related tasks for a sick or disabled family member compared to young men in line with research by Bowlby et al (2010), Evans and Becker (2009) and Milligan (2000). However, young men also had considerable caring responsibilities, although the nature of their caregiving was more focused on providing financial support to their families (Day and Evans 2015). Both of these types of familial caregiving responsibilities impacted on young people's ability to complete socially expected transitions to adulthood, especially education and employment, which are regarded as increasingly important in a globalising world. Those who were not caring for someone sick or disabled, however, still had significant responsibilities at home, often to provide for older and younger relatives, such as siblings and grandparents. This further demonstrated how young people's lives should be more appropriately considered as being part of a continuum of care as they negotiate their transitions to adulthood alongside a range of caregiving activities.

While a number of additional factors impacted on the young people's transitions to adulthood – poverty and lack of educational opportunity being particularly key – caring responsibilities were also an influence. Young people described choosing to delay completing certain transitions to adulthood, such as marriage or migrating to a town or city, in order to remain taking care of their families. This demonstrated the continuum of care and associated responsibilities took place within all four phases of the ethic of care (Tronto 1993). Caregiving was not just the physical act of looking after someone, but taking responsibility for others by providing for them and caring about their health and wellbeing generally, often over and above the young people's own needs. Young people still aspired to reach adult status, to complete their education, find secure paid employment and have a home, spouse and children, but realised they had to balance these aspirations with their caregiving and family responsibilities.

Conclusion

This chapter discusses how caregiving in Zambia is rarely recognised as an issue affecting Zambian youth, or seen as being over and above the responsibilities young people are expected to hold within the household. Young people in Zambia, in common with youth in other eastern and southern African countries severely affected by the HIV epidemic (Skovdal 2011, Evans and Becker, 2009, Robson et al 2006), are taking on increasing caring responsibilities for parents and relatives living with HIV, as well as for those with disabilities and other chronic illnesses. The research demonstrated that there was still an underlying expectation that care would be undertaken by 'adult' family members. This was despite acknowledgement that household structures are changing, HIV has had a significant impact on the 'middle generation' and young people are increasingly heading households consisting of young siblings or elderly grandparents, who may require physical care as well as financial support. Where concern was raised about young people's caregiving responsibilities, it focused on 'children' in line with global notions of childhood as a time of innocence free from adult responsibility (James et al 1998). This was despite recognition that young people in Zambia, as well as many other African communities, take on an increasing number of household responsibilities as they grow from children to young adults in preparation for one day having their own homes, as well as in support of the family (Solberg 1997, Punch 2001, Katz 1993, 2004, Miller 2005, Evans and Becker 2009). Youth were assumed, because of their age, to have developed a wider skill set and network of resources that would help them to cope with their caregiving responsibilities better. Young people often saw their caring role as 'just part of what they do' as part of their family obligation, and therefore not worthy of additional attention. What was evident however, was that caring responsibilities, combined with growing up in a compromised socioeconomic and political context, meant that few young people were successfully completing important transitions to adulthood, such as completing education and finding employment (Day and Evans 2015). It was also evident that caring was something that young people moved in and out of and had influenced all the young people's lives and experiences at one time or another as they had grown up. The research therefore calls for further exploration of the wider nuances present in young people's lives in relation to the continuums of care that young people move in and out of. A more relational conceptualisation of young people's life transitions that are characterised by interdependence between family members and the 'continuum of care' that shapes young people's transitions to adulthood is therefore needed. This may be better addressed by adopting a 'livelihoods' approach through which to analyse the young people's lives. Within this study all the young people were negotiating their lives against a wider background of poverty, which made them more vulnerable to acquiring caregiving roles in the first place, as well as impacting on their transitions to adulthood. A livelihoods approach would allow a more detailed analysis of the effect of poverty on young people's lives to take place, particularly looking at the cause or effect of poverty on the 'critical moments' in young people's lives. The research also calls for government agencies and NGOs to recognise the increasing number of 'youth' who are taking on significant caring responsibilities within their families and recognise the specific circumstances, vulnerabilities and capabilities of this increasing population. An understanding of what level of care and responsibility is acceptable without wider support mechanisms in place needs to be established in order to target appropriate responses when caring becomes over and above young people's capabilities. A greater recognition of the types of caring roles young people are undertaking and how this may impact on their ability to complete their socially expected transitions to adulthood needs to be established and measures put in place to support young people struggling with the burden of care and allow them to focus on the completion of education and other transitions to adulthood, including finding employment, which will have wider benefits for the individual and their family in the long term.

Acknowledgements

Thanks go to the University of Reading who provided me with a PhD studentship and the Royal Geographical Society (with the Institute of British Geographers) Slawson Award for financial support for the research.

References

Aldridge, J. and Becker, S. (1993) *Children who care: inside the world of young carers*. Loughborough, Loughborough University.

Ansell, N. (2005) *Children, youth and international development*. London, Routledge.

Barry, M. (2011) 'I realised that I wasn't alone': the views and experiences of young carers from a social capital perspective. *Journal of Youth Studies* 15, pp.523–39.

Bauman L.J., Foster, G., Johnson Silver, E., Berman, R., Gamble, I., and Muchaneta, L. (2006) Children caring for their ill parents with HIV/AIDS. *Vulnerable Children and Youth Studies* 1(1), pp.56–70.

Becker, S. (2007) Global perspectives on children's unpaid caregiving in the family: research and policy on 'Young Carers' in the UK, Australia, the USA and Sub-Saharan Africa. *Global Social Policy* 7, pp.23–50.

Bourdillon, M. (ed.) (2000) *Earning a life: working children in Zimbabwe*. Harare, Weaver Press.

Bowlby, S. (2012) Recognising the time-space dimensions of care: caringscapes and carescapes. *Environment and Planning A* 44, pp.2101–18.

Bowlby, S., Gregory, S. and McKie, L. (1997) Doing home: patriarchy, caring and space. *Women's Studies International Forum* 20, pp.343–50.

Bowlby, S., McKie, L., Gregory, S. and MacPherson, I. (2010) *Interdependency and care over the lifecourse*. London, Routledge.

Collard, D. (2000) Generational transfers and the generational bargain. *Journal of International Development* 12, pp.453–62.

Day, C and Evans, R. (2015) Caring responsibilities, change and transitions in young people's family lives in Zambia. *Journal of Comparative Family Studies* 46, pp.137–52.

Dearden, C. and Becker, S. (2000) *Growing up caring: vulnerability and transition to adulthood – young carers' experiences*. York, Joseph Rowntree Foundation.

Evans, R. (2010) Children's caring roles and responsibilities within the family in Africa. Geography Compass, 4 (10). pp. 1477–1496.

Evans, R. (2012a) Sibling caringscapes: time-space practices of caring within youth-headed households in Tanzania and Uganda. *Geoforum* 43, pp.824–35.

Evans, R. (2012b) Safeguarding inheritance and enhancing the resilience of orphaned young people living in child-and youth-headed households in Tanzania and Uganda. *African Journal of AIDS Research* 11, pp.177–89.

Evans, R. (2013) Young people's caring relations and transitions within families affected by HIV. In Ribbens McCarthy, J., Hooper, C.A. and Gillies, V. (eds.) *Family troubles? Exploring changes and challenges in the family lives of children and young people.* Bristol, Policy Press, pp.233–43.

Evans, R. and Becker, S. (2009) *Children caring for parents with HIV and AIDS: global issues and policy responses.* Bristol, Policy Press

Evans, R. and Day, C (2011) Inheritance, poverty and HIV/AIDS: Experiences of widows and orphaned youth heading households in Tanzania and Uganda. Manchester, Chronic Poverty Research Centre.

Foster, G., Levine, C. and Williamson, J. (eds.) (2005) *A generation at risk: the global impact of HIV/AIDS on orphans and vulnerable children.* Cambridge, Cambridge University Press.

James A., Jenks C. and Prout, A. (1998) *Theorizing childhood.* Cambridge, Polity.

Katz, C. (1993) Growing circles/closing circles: limits on the spaces of knowing in rural Sudan and US cities. In Katz, C. and Monk, J. (eds.) *Full circles: geographies of women over the life course.* London, Routledge, pp.88–106.

Katz, C. (2004) *Growing up global: economic restructuring and children's everyday lives.* Minneapolis, University of Minnesota Press.

Laird, S. (2005) International child welfare: deconstructing UNICEF's country programmes. *Social Policy and Society* 4, pp.457–66.

Lawson, V. (2007) Geographies of care and responsibility. *Annals of the Association of American Geographers* 97, pp.1–11.

Miller, P. (2005) Useful and priceless children in contemporary welfare states. *Social Politics* 12, pp.1–39.

Milligan, C. (2000) 'Bearing the burden': towards a restructured geography of caring. *Area* 32, pp.49–58.

Milligan, C. (2005) From home to 'home': situating emotions within the caregiving experience. *Environment and Planning A* 37, pp.2105–20.

Ministry of Sport Youth and Child Development (MSYCD) (2006) *National Youth Policy, 2006.* Lusaka, MSYCD.

Muyinda, H., Seeley, J., Pickering, H. and Barton, T. (1997) Social aspects of AIDS-related stigma in rural Uganda. *Health and Place* 3, pp.143–7.

Ogden, J., Esim, S., and Grown, C. (2006) Expanding the care continuum for HIV/AIDS: bringing carers into focus. *Health, Policy and Planning* 21, pp.333–42.

Payne, R. (2012) Agents of support: intra-generational relationships and the role of agency in the support networks of child-headed households in Zambia. *Children's Geographies* 10, pp.293–306.

Potter, R., Conway, D., Evans, R. and Lloyd-Evans, S. (2012) *Key concepts in development geography.* London, Sage.

Punch, S. (2001) Household division of labour: generation, gender, age, birth order and sibling composition. *Work, Employment and Society* 15, pp.803–23.

Robson, E. (2004) Hidden child workers: young carers in Zimbabwe. *Antipode* 36, pp.227–48.

Robson, E., Ansell, N., Huber, U.S., Gould, W.T.S. and van Blerk, L. (2006) Young caregivers in the context of the HIV/AIDS pandemic in sub-Saharan Africa. *Population, Space and Place* 12, pp.93–111.

Rose, H. and Bruce, E. (1995) Mutual care but differential esteem: caring between older couples. In Arber, S. and Ginn, J. (eds.) *Connecting gender and ageing: a sociological approach.* Buckingham, Open University Press, pp.114–28.

Sevenhuijsen, S. (1998) *Citizenship and the ethics of care: feminist considerations on justice, morality and politics*. London, Routledge.

Skovdal, M. (2011) Examining the trajectories of children providing care for adults in rural Kenya: Implications for service delivery. *Children and Youth Services Review* 33, pp.1262–69.

Smyth, C. Blaxland, M. and Cass, B. (2011) So that's how I found out I was a young carer and that I actually had been a carer most of my life. *Journal of Youth Studies,* 14(2) pp.145–160.

Solberg, A. (1997) Negotiating childhood: changing constructions of age for Norwegian children. In James, A. and Prout, A. (eds.) *Constructing and reconstructing childhood: contemporary issues in the sociological study of childhood*. London, Routledge, pp.118–27.

Thomson, R., Bell, R. Holland, J., Henderson, S., McGrellis, S. and Sharpe, S. (2002) Critical moments: choice, chance and opportunity in young people's narratives of transition. *Sociology* 36, pp.335–54.

Tronto, J. (1993) *Moral boundaries: a political argument for an ethic of care*. New York, Routledge.

United Nations (2007) Definition of Youth. Available at www.un.org/esa/socdev/documents/youth/fact-sheets/youth-definition.pdf (accessed 22 Feb 2017).

van Blerk, L. and Ansell, N. (2007) Alternative care giving in the context of AIDS in southern Africa: complex strategies for care. *Journal of International Development* 19, pp.865–84.

Wardale, L. (2003) *Barnardo's action with young carers and keeping the family in mind*. Liverpool, Barnardo's.

Worth, N. (2009) Understanding youth transition as 'becoming': identity, time and futurity. *Geoforum* 40, pp.1050–60.

10 Caring after parental death

Sibling practices and continuing bonds

Ruth Evans

Introduction

Family geographies are still in their infancy and very little geographical work, and social science research more broadly, has investigated sibling relations to date. Geographies of care have sought to understand the complexity of caring relations within families (Evans and Thomas 2009, Hall 2016) and there is growing interest in the spatialities of intergenerational relations (Vanderbeck and Worth 2015). As I have argued elsewhere (Evans 2015a), however, an intergenerational lens should not lead to a neglect of *intra*-generational relations, since siblings and peers often have a significant influence on young people's lives. The often reciprocal nature of sibling relations, and their continuity and longevity throughout the lifecourse make sibling practices particularly worthy of attention.

A small body of sociological work discusses the influence of gender and birth order on children's sibling relations in the global North (Brannen et al 2000, Mauthner 2005, Edwards et al 2006, Punch 2008, Klett-Davies 2008). Literature that explicitly addresses sibling caregiving (Kosonen 1996) often discusses situations where siblings have a need for care that is related to childhood disability (Stalker and Connors 2004, Burke 2004) or older age (Brannen et al 2000, Barnes 2012). While some anthropological work discusses 'sibling caretaking' and childcare practices (Cicirelli 1994, LeVine et al 1996, Rabain-Jamin et al 2003), development and geographical research on care in Africa tends to be dominated by studies of orphanhood and the HIV epidemic and yet rarely discusses young people's specific experiences of bereavement (Nyambedha et al 2003, Skovdal et al 2009, Abebe 2012).

This chapter seeks to respond to these research gaps. While my previous work has analysed the time-space dimensions of caring practices in sibling-headed households in Tanzania and Uganda (Evans 2012a) and changing intergenerational caring relations in bereaved families in Senegal (Evans 2014), this chapter brings together common themes emerging from these two qualitative studies. I draw on the concepts of 'sibling practices', 'displaying family' and 'continuing bonds' to analyse sibling caring practices and relations with the deceased. I first briefly outline the research studies on which this chapter draws. I then discuss the literature and my conceptual approach, before analysing the empirical findings.

The first study focused on sibling care in Tanzania and Uganda (2008–9). The discussion of the findings draws predominantly on semi-structured interviews conducted with 14 young people (9 girls, 5 boys, aged 12–23) orphaned by AIDS living in 11 sibling-headed households in urban and rural areas in Tanzania and Uganda and focus groups with a further 15 young people. Participatory feedback workshops were also held with 33 young people (15 siblings heading households who had been interviewed and 18 of their younger siblings) a year after the initial fieldwork. See Evans (2011, 2012a and b) for more details.

The second study explored inheritance, poverty and family relations in Senegal, West Africa (2011–12), a context with relatively low HIV-prevalence compared to Eastern and Southern Africa. The discussion draws predominantly on semi-structured interviews conducted with 5 young women and 5 young men (aged 15–27 years) whose mother/father or both had died in recent years, living in two rural communities and in low-income neighbourhoods of Dakar, the capital. See Evans (2014, 2016) for more details.

In both studies, audio-recorded interviews and focus groups were transcribed/translated with research assistance (from Kiswahili into English for the Tanzanian fieldwork, transcribed in English for the Ugandan fieldwork and transcribed in French or translated from Wolof or Serer into French for the Senegal study). For the Senegalese data, I translated quotations used in this chapter into English from the French transcripts. Both projects were granted ethical clearance by the University of Reading Research Ethics Committee. Participants' accounts have been anonymised throughout the chapter to protect their identities.

Bereavement and sibling relations

Although geographical research about bereavement and mourning in the global South is limited to date, the importance of understanding the relationships between space/place and death has been increasingly recognised (Hockey et al 2010, Maddrell and Sidaway 2010). Recent literature has explored care of the dead, as well as care of the living following a death (Ribbens McCarthy 2012, Evans et al 2016) and argued that relationships with the dead are often characterised by continuing bonds (Klass et al 1996) and a dynamic relational notion of 'absence-presence' (Maddrell 2013). As Ribbens McCarthy (2009, p.23) notes, "the social realities of death and bereavement for young people will vary significantly across time and place". My research in Senegal suggests that parental death represented a 'vital conjuncture' for young people which could reconfigure intergenerational caring relations and lead to shifts in power dynamics, as eldest sons asserted their position of authority and inherited a greater share of their father's assets than their sisters (Evans 2014).

Few studies to date have explored sibling relations following parental death in the global South. Child development and anthropological research has highlighted the cultural significance of older siblings' roles in caring for younger siblings and in socialisation and informal training in many African societies and in the global South in general (Weisner 1982, Cicirelli 1994, LeVine et al 1996, Rabain-Jamin et al 2003). 'Sibling caretaking' of young children by older siblings is commonly used as an important childcare strategy to free up time for mothers to engage in other productive or reproductive tasks, as well as forming a crucial element in children's socialisation in familial responsibilities and cultural values. Both older and younger siblings engage in 'caregiving' and 'care-receiving' roles within the family and interdependent sibling relationships, particularly brother-sister relationships, continue throughout the lifecourse in many societies in Sub-Saharan Africa (Weisner 1982, Rabain-Jamin et al 2003). Furthermore, childcare of younger siblings often forms part of the continuum of young caregiving in households affected by HIV and other chronic illnesses in African contexts (Evans 2010).

Some studies have explored the circumstances of child- and youth-headed households in the global South, where siblings live independently in 'self-care arrangements' without adult supervision (Thurman et al 2006, Dahlblom 2008, Payne 2009, Meintjes et al 2010). Young people's resilience and agency in adopting coping strategies is acknowledged, alongside the poverty, stigma and marginalisation they may face (Thurman et al 2006, Francis-Chizororo 2008). Despite the cultural significance of sibling caregiving, Kesby et al (2006) suggest that the situation of orphaned children growing up in child-headed households in the context of the HIV epidemic challenges local understandings of childhood as well as universal models. From the commencement of their caring trajectories, young people are engaged in difficult changes in their family lives, taking on 'adult' childcare roles and responsibilities that do not conform to Western or African constructions of childhood (Evans 2011).

Although research about child- and youth-headed households in Eastern and Southern Africa usually concerns orphaned young people, these studies rarely focus on experiences of 'bereavement' *per se*. This reflects a broader trend in development-oriented research, which focuses more on the material and social impacts of orphanhood in the context of the HIV epidemic, rather than on young people's emotional responses to parental death. Studies which do investigate the emotional effects of parental HIV-related sickness, bereavement and caring on children (Bauman et al 2006, Cluver et al 2007) tend to adopt a medicalised/psychological framework. My earlier work developed the notion of 'sibling caringscapes' to conceptualise the time-space patterning of sibling care, integrating young people's everyday caring activities with the emotional geographies and reflective positions that young people adopt and negotiate across different locales through time (Evans 2012a). The competing rhythms of bodies, schooling, work and seasonal agricultural production among young people heading households resulted in 'arrhythmia' and time scarcity, which had detrimental effects on young people's health, education, future employment prospects and mobility.

Conceptual approach

Edwards et al's (2006) concept of 'sibling practices', based on Morgan's (1996, 2011) concept of 'family practices', is useful in highlighting the dynamics of sibling interactions following parental death. Family practices are "practices that matter to the persons concerned as having something to do with 'being a family'" (Edwards et al 2006, p.60). 'Sibling practices' draw attention to the ways that sibling relationships are socially constructed and "enacted as part of everyday life together, creating meaning, rather than enshrined in biological fact and delineated by universal developmental processes" (ibid., p.61). Furthermore, the idea of practices denotes a sense of fluidity, as these practices are constructed and negotiated between siblings and interact with wider systems of meaning and power relations such as gender, age and status hierarchies (ibid.). Building on 'family practices', Finch (2007, p.66) argues that "families need to be 'displayed' as well as 'done'". The notion of 'displaying' emphasises "the fundamentally social nature of family practices, where the meaning of one's actions has to be both conveyed to and understood by relevant others if those actions are to be effective as constituting 'family' practices" (ibid.).

Alongside these understandings of 'sibling practices' and 'displaying family', I draw on death studies literature which has theorised the relationships between the living and dead in terms of 'continuing bonds' (Klass et al 1996). Such bonds are shaped by relationships in life, by the nature of the death itself and by other social factors such as economic status, ethnicity, religion, age, gender and sexuality (Howarth 2007, Ribbens McCarthy 2012) and place (Maddrell 2013). This chapter draws on these understandings of the social and emotional aspects of 'bereavement' through a focus on relations among surviving family members and on young people's material and non-material inheritance and continuing bonds with the deceased.

Young people's responses to death

In the Senegal study, most of the young people interviewed expressed profound feelings of loss and sadness about the death of their parent. Young people often remembered clearly the day their parent had died and the moment they had received the news. Aita, a young woman (aged 20) living in Dakar, who was visibly upset during the interview and gave very brief responses, expressed how difficult it was coping with the sense of solitude and poverty that she, her mother and siblings faced at home since her father's death. She had recently started a tailoring apprenticeship, but had worked for several years as a domestic worker to help her mother support the family. When asked about her future aspirations, Aita simply said she wanted to work and even if she married, she wanted to remain close to her mother. She appeared to have little hope for the future, aside from her religious faith: "The problems will never end. […] God is merciful, he will get us out of these problems."

In another family experiencing chronic poverty, material changes following the death were bound up with the emotional impacts. Khady, a young woman (aged 20) who was the eldest sibling living at home in a rural area and whose parents had died within a fortnight of each other three years before the interview, explained how the deaths had affected her: "The loss of my parents affected me a lot […] At that time, I was in Troisième [equivalent of Year 10 in UK secondary school] I failed […] I was thinking of them when I was in class." Her older brother, the eldest sibling, had migrated to Europe in search of work, due to their precarious situation after her parents' death, and at the time of interview was reported to be held in a detention centre there due to a lack of papers. Khady commented on how her relationships with her younger siblings had changed, since they had moved to live with her grandmother in the village: "We have become much closer, sometimes when you see them, you feel that they are missing something [...] You see that they are missing something." Thus, while becoming closer to surviving family members, the siblings simultaneously continued to feel the absence of their parents, manifested as a visible 'lack' or as something missing in their lives.

Thus, negative changes in families' material circumstances were often entwined with young people's emotional responses, as has been confirmed by our more recent research in Senegal (Evans et al 2016). Indeed, writing of HIV-affected households in South Africa, Demmer (2007) suggests that struggles for daily survival may take precedence over the expression of grief. Alongside material differences, age-based differences were also evident in young people's responses to parental death in Senegal, although the research was based on a very small sample size. Younger participants (aged 15–20) appeared to experience a deeper level of poverty and greater social and emotional disruptions following parental death, compared to older youth (aged 25–30) interviewed.

Since the focus of the study in Tanzania and Uganda was on sibling care, young people's narratives of events following their parent's death focused predominantly on their transition into caring, including deciding whether to stay together with their siblings in the parental home or to live with foster relatives for a period. When prompted, Rickson described his feelings in the aftermath of his parents' deaths and highlighted the importance of peer support in helping him to adjust to living alone with his siblings:

> I felt very bad because of losing both parents, I was sad for about four months. I was thinking of my parents. But I had to get used [to it] and leave it, and be like others, because others lose their parents as well. Therefore we were exchanging ideas with other youths like me who have lost their parents, on how to live, we exchanged ideas and we therefore removed that idea of saying that we have lost our parents. We got used to it.

Many of the eldest siblings heading households felt that they had developed life skills in managing a household and learned to become more responsible and mature, which appeared to foster their resilience. As Amina commented (via the translator): "She has learned to be responsible, so she feels she can make it in life, regardless, whether bad or good."

In addition to commenting on negative effects on their lives, older youth (aged 25–30) in Senegal also identified positive changes associated with their parent's death, engaging in 'meaning-making' to reframe the death (Ribbens McCarthy 2009). Badara, a young man (aged 26) living in an extended family household headed by his uncle in a rural area who was studying at university in the city, commented on how difficult it was trying to replace his father's income and role as head of the household. Badara also suggested, however, that the deaths of his parents and his religious faith had made him more determined to succeed in his education and make something of his life:

> it's these two events [the death of his mother and his father] which really marked me the most and have changed many things, because after the death, I had courage because I had to work to look after my brothers and after a while, I said to myself, it's God who wanted it to be this way and I had the courage to learn to be someone who achieves things in life.

Similarly, Diana, a young woman (aged 27), identified positive changes in her own personal development following her father's death. Diana, her mother and siblings experienced considerable poverty, were forced to relocate to another rented property and did not receive their share of inherited assets from her father's first wife and children (her mother was the second wife in a polygamous union) until several years later. Nevertheless, Diana felt that she had developed in terms of her maturity and sense of responsibility for her younger siblings:

> I became more aware of the situation, braver, it's like I matured 10 years, because with papa, we were spoilt too much, every evening, he brought us oranges, we all sat around him. But after his death, I felt that I had grown up, I see my brothers and my sisters differently [...] because since his death, it's like I am responsible for my younger brothers and sisters.

Parental death thus appeared to re-affirm the familial obligations of older youth to provide for surviving parents and younger siblings as part of the conventional 'generational bargain' (Collard 2000).

Sibling practices

Anthropological research has highlighted the ways that parenting cultures and childcare disciplining methods are culturally determined (LeVine et al 1996, Weisner 1982). In both studies, as noted above, the death of a parent reaffirmed the conventional role that older siblings were expected to play in providing informal teaching and setting a good example to their younger siblings. Sibling birth order and age-based hierarchies were significant in determining the differing roles undertaken by young people (Punch 2001, Rabain-Jamin et al 2003, Evans 2012a). Furthermore, as Carsten (2004, p.35) argues, 'kinship is made in houses through intimate sharing of space, food, and nurturance that goes on within domestic space'. Siblings thus made and re-made family relations and displayed 'family' through shared 'sibling practices' in the (inherited) parental home.

Most young people interviewed in Tanzania and Uganda felt that they had a close, loving relationship with their younger siblings and that their siblings respected them, citing these as some of the best aspects of caring for their siblings. Many young people heading households felt that their role was to set a good example to their younger siblings through their own behaviour and to provide moral guidance and informal teaching, as their mother or father would have, had they been alive. Sophia (aged 21, Dar es Salaam) who cared for her young child and three siblings, one of whom had a mobility impairment, said:

> I am supposed to lead my siblings in a proper life. When they do wrong, I have to be the first to correct them. I also should struggle to avoid doing what is not good so that they should not copy bad things from me, they should copy good things.

The need to 'display family' and demonstrate good moral values and behaviour may be especially important for child- and youth-headed households, given the wider socio-cultural context in which such households were considered to be deviant. Such perceptions were manifested in the suspicion and harassment that young people experienced from neighbours and community members (Evans 2011), which may foster a greater sense of solidarity and self-reliance as a sibling group. For example, when asked "who can you talk to if you have a problem or you're worried or upset about something?", Godfrey (aged 17, rural area, Uganda) replied (via translator), "now he just keeps quiet, no one [...], they decided to finish their problems here, with [among] the brothers".

In the Mbeya workshop, both younger and older siblings decided that their key message for local leaders, community members and NGO staff which they painted on their art poster should be, "We are happy living together as a family" (see Figure 10.1). This suggests a strong desire to 'display' the way that they were doing 'family' to others in order to achieve acceptance and respect in the community. Indeed, older siblings in the Mbeya workshop identified '*being respected within and outside the household*' as one of the best aspects of caring for their siblings.

Some young people's experiences reveal a fluidity of caregiving roles and the constantly shifting nature of sibling practices which change according to age, competencies and circumstances. Sophia said: "My position does change as I grow up, because my roles change too day by day." Many young people regarded themselves as adults/parents in terms of their responsibilities to provide for the needs of their younger siblings. Rickson (aged 19, rural area, Tanzania) explained his role in advising and teaching his younger siblings, considering himself a 'parent' and 'grown up':

> Because I'm the eldest, I have to correct them when they make mistakes. [...] at least I am a grown up and I know what life is like. But when my younger brother does a mistake, I have to correct him, and show him a good way so that he won't be lost in his future. So I advise him, but sometimes I have other work and I have my studies. I have to plan what should be done, because they are still young, they cannot plan well.

Some young men appeared to subscribe to a particularly masculine notion of their caring role as the male head of the household, an authority figure who leads the

Figure 10.1 Young people's life-size art poster created in participatory feedback workshops in Mbeya, Tanzania, with their key message in Kiswahili: "We are happy living together as a family" (Evans 2011)

family and provides physical protection from threat and violence (Edwards et al 2006). When asked how they saw their position in the family, one young man in a focus group in Tanzania commented:

> This is the position that is usually taken by the father. This position pleases me because I am the protector of the house. This is his position. If there are problems, for example with the police coming to the house – I don't know, [say] something has happened – then my position is to protect the house.

Young men often asserted their role as the male head of household and a figure of authority in disciplining their younger siblings.

Although sibling relations were not a key focus of interviews in Senegal, some young men also reported reasserting their role in teaching and disciplining their younger siblings following parental death. Anthropological research with Wolof siblings in Senegal emphasises older siblings' teaching and caregiving roles; older siblings not only give orders and thereby grant younger siblings responsibilities and allow them to participate, but this context also facilitates dialogue and language learning (Rabain-Jamin et al 2003). Samba, a young man (aged 27) who lived with his father, his father's new wife and his younger siblings in a rural area, considered that he was more responsible for his siblings' informal education since his mother had died:

> I'm the one who looks after their education. It's enough to call me for them to do it. [...] it is enough to use a certain language, they will respect you, there are words that if you address them to someone, they will grant you respect. It's not only the cane which educates. Here, I am the one who is responsible, because it's me who dictates what should be done, they carry it out, I order them to not do it, they don't do anything.

This suggests that a parent's death may reinforce the eldest male sibling's position of authority over younger siblings through 'family practices' and everyday routines within the home (Evans 2014).

Omari and Mbilinyi (1997, p.48) suggest that corporal punishment is enshrined in African culture and traditions, pointing out, "Some people in the society, especially those who consider themselves as custodians of societal cultures, are blind towards the negative impact of what has been upheld by the previous generation and perceived as good". Corporal punishment as a means of disciplining children is maintained by age hierarchies and the exploitation of power imbalances between children and parents/adult relatives and older siblings (Evans 2004). Within this cultural context, it is unsurprising that some siblings heading households said that they used a cane to discipline their younger siblings when they did not respect their authority. For example, when Harriet (aged 17, rural area, Uganda) was asked how she got on with her younger siblings, she replied (via the translator): "The boys used to fight but she caned them, she disciplines them, they no longer fight. Now they are good, they listen to her, they respect her."

However, several young people said that they resolved conflicts with their siblings through more democratic disciplining methods, such as through dialogue and giving advice. Hamisa commented that she used the same disciplining methods with her young cousin that her mother had used with her, "I call her, I discuss with her just like our mother did. I tell her that 'If you do not respect me, who will you respect?'" Isack (aged 12, rural area, Uganda) said that one of the best things about living with his older brothers was that they discussed ideas together and advised each other:

> Even we discuss ideas, if one does a bad thing, then we correct him ... even not going at school, we say that if you don't go to school now, in the future you will face problems. We advise each other.

Older siblings thus take over the deceased parent's role in providing advice to younger siblings.

Children's misbehaviour and refusal to obey their elder sibling, however, could cause tensions and frustrations for young people heading households.

Rickson said that the worst aspect of caring for his siblings was when his younger brother had not undertaken the household tasks he had asked him to do:

> if I tell [younger brother] that when you come from school you should go to fetch water, you come and find him playing and he has not done what I asked him, I become very angry and say, why didn't he do what I told him? … if it was his father he would have respected him.

He felt that he was not granted the respect that his father would have commanded, had he still been alive. In a focus group in Mbeya, Tanzania, some young women discussed how tensions with their younger siblings made them feel unvalued, despite the care they provided:

> I don't like it when my younger siblings insult me and when they treat me as if they hate me, I feel really bad. At times like that I feel as if they don't appreciate me at all and that I would be better off just to leave them. But then people would say "Hey, she left her younger siblings all on their own".

This highlights the difficult emotional dimensions of sibling caregiving. Younger siblings also commented on the emotional impact of conflict with older siblings, saying in the Kagera workshop, they felt bad if they were "harassed by their older brothers and sisters", were "beaten by their brother or sister for no reason", were "given too much work" or were "insulted by a guardian and they ended up hating each other". Some young people in Tanzania appreciated being able to ask NGO workers for advice and support in resolving disputes with their younger siblings.

While some young people in Tanzania and Uganda were proud of their caring role as head of the family, several highlighted the loneliness of their parental role, missing the love, advice and guidance that their late parents had provided. Recent research in Senegal also suggests that 'bereaved' young people often missed the advice their parent used to provide, implying that no one else was able to replace this parental role in the same way (Evans et al 2016). Furthermore, although younger siblings shared the domestic responsibilities in sibling-headed households, emotional support provided by older siblings was not necessarily reciprocated. Young people struggled to provide for their younger siblings' material and emotional needs, while their own emotional needs for love, support and guidance often went unnoticed. When asked how he felt about looking after his siblings when his mother left and later died, Godfrey's (aged 17, rural area, Uganda) response (via the interpreter) was:

> He feels pain, because he's not the mother, he's not the father, but he have to take the responsibility, especially on clothing, mm, looking for eats [food], as if he's a parent, the one who doesn't go now to school.

Godfrey felt that he had to act as a father to his younger brothers, even though he felt he was still a young person, and was forced by circumstances to take on this position.

In the Kagera workshop, eldest siblings identified a number of aspects of their role that made them feel bad, including "worrying about my competence" [to meet their needs], being "forced to be happy all the time although I feel sad", "feeling lonely when I see my friends being brought up by their parents", and "carrying the responsibility of a grown-up when I am still a child". When prompted further, older siblings said that they did not want to share their thoughts and emotions with their younger siblings, due to a desire to protect them from emotional distress, as has been reported elsewhere of caring relations in families affected by HIV (Evans and Thomas 2009).

Continuing bonds

Although the studies reported here did not have an explicit focus on care of the dead or continuing bonds, young people's narratives demonstrate that continuing the legacy of their deceased parent was important to them. Young people in Senegal, Tanzania and Uganda were concerned to fulfil their deceased parents' wishes and continue to care for siblings and other relatives and manage the household in the same way that their parent had. Older siblings' concern to set a good example to their younger siblings, in their parent's absence, may be linked to a desire to demonstrate the values and behaviour they had inherited from their parent, as a means of expressing continuing bonds and 'displaying' family and the legacy of their deceased parent both to each other and to those outside the household.

In the Tanzania and Uganda research, several young people often regarded their caring role for their siblings as part of their commitment to fulfil their deceased parent's wishes. Parents living with HIV-related illness sometimes sought to prepare their children for their death, expressing their wish that the siblings remained together to ensure their survival. As Amina (aged 16) caring for four younger siblings, three of whom were living with HIV, in Kampala commented: "When she's not yet dead [before her death], when she told me, I agreed to stay with my brothers and sisters, because my mother was not in good health, because she was sick." Amina regarded fulfilling her deceased mother's wishes as the best aspect of caring for her siblings (via the translator): "She likes it because she did not disappoint the late mother, that's the best thing, because she does it proudly and willingly, even though it sometimes becomes difficult."

In the Senegal research, following his father's and his uncle's deaths, Issa, a young man (aged 25) living in a rural area, considered himself the head of household and *kilifa* [Wolof for head of the family, moral authority and decision-maker]. His father had entrusted his younger siblings to his care before his death. Issa expressed a strong sense of duty and responsibility to manage the household in the same way as his father, providing for his widowed stepmother and younger siblings, as a means of reciprocating the care his father had formerly provided, as part of the 'generational bargain' (Collard 2000):

> This doesn't cause me any problems. This is down to the will of God. Just, he [his father] did everything he could do for me. He did many things for me.

> Now that he's no longer there, I look after everything. [...] I do as he used to do at the time. I continue what my father used to do.

In the East African study, some young people expressed a strong sense of belonging to the home where they had lived with their parent(s) and were concerned to stay there with their siblings in order to safeguard their inherited assets from unscrupulous relatives or neighbours. In rural areas, the main assets young people inherited were the house, farmland, food stores and agricultural equipment, which they felt were at risk of property grabbing if they moved away to live with foster relatives (see Evans 2012b, 2015b for more in-depth discussion of disinheritance).

Deceased parents' graves were usually located in the *shamba* (Kiswahili for garden/field) outside the house, as is customary in the research locations in Tanzania and Uganda, which contrasts to the research from Senegal where the dead were buried in village or town cemeteries. Young people in Tanzania and Uganda showed me their parents' graves when I conducted interviews at their homes. Physical proximity to the grave and the buried remains of their deceased parent may reinforce young people's sense of belonging to their inherited parental home. Maintaining and weeding the grave and continuing to cultivate the land they had inherited can be seen as an important means of young people caring for the dead and expressing continuing bonds.

In the Senegal research, young men in rural areas reported a heightened sense of responsibility to work hard to cultivate crops on the land that they had inherited and to manage other assets such as livestock, in addition to studying hard among those attending school or university. Young men's responsibilities to cultivate the land were closely bound up with their deceased father's identity and a sense of belonging (Evans 2014). Abdoulaye (aged 19 years) worried about how to make good use of his father's large fields, which he and his brother had inherited but were not currently able to cultivate because Abdoulaye lacked sufficient labour resources and the means to pay labourers to work on the land. He was concerned about leaving his father's land untended, as it would be at risk of reallocation by the rural council (Evans 2014). Daughters did not usually inherit land, except for any rice paddies their mother had cultivated in one of the rural communities, and according to Islamic law, usually received half the share of other assets or belongings that their brothers inherited (Evans 2016).

Some young people in Senegal commented on non-material aspects of inheritance and the cultural transmission of values they felt their parent had passed on to them, such as a commitment to work hard in their studies. Diana (aged 27) and her sister cited obtaining their Baccalaureat (equivalent of A-level qualifications) as critical moments (Thompson et al 2002) in their lives due to their father's aspirations for their educational success:

> because for us it's a contract that we have fulfilled because our father held our studies in high regard, so that when we succeeded, we said to ourselves, "he must be pleased, there where he is".

Young people thus sought to fulfil their parents' wishes and aspirations for their future, as a means of expressing their continuing bonds with the deceased and perhaps 'displaying' their parent's legacy to others.

For one young person, the legacy of the deceased was evidenced through the strengthening of existing kinship relations with surviving extended family members. Despite studying in the city, Badara nevertheless imagined living in the village near his relatives in future:

> Me, I prefer to live here next to my relatives, because if you have lost a mother or a father, you will still have close relatives, because your mother had brothers and sisters, if she died, they are there, her sisters become your mothers, and her brothers, your uncles and the same for your father. They are not the same, but they are your parents/relatives.

The French word '*parents*' used at the end of this extract means 'relatives' as well as 'parents' in English; the phrase '*liens de parenté*' is commonly translated as 'kinship ties'. Thus, maintaining close ties with extended family members appeared to be important in enabling Badara to express continuing bonds with his deceased parents, since his uncles and aunts were regarded as 'representatives' of his deceased mother and father (Evans 2014).

Conclusion

The research findings from Tanzania, Uganda and Senegal suggest that young people respond in different ways to the death of a parent. Continuing feelings of grief and emotional upheaval were entwined with significant changes in young people's material and social circumstances, which they may find difficult to adjust to. Alongside experiences of poverty and other negative impacts, however, eldest siblings heading households in Tanzania and Uganda and older youth in Senegal also engaged in 'meaning-making' (Ribbens McCarthy 2009) by identifying positive aspects of their parent's death and caring roles, such as becoming more mature, feeling a greater sense of responsibility for their siblings and developing important life skills.

The notion of sibling practices (Edwards et al 2006) emphasises the ways that sibling relationships are socially constructed and performed through everyday activities and routines and often linked to gendered and intra-generational birth order hierarchies. The findings reveal the fluidity of roles and diverse sibling practices that young people engage in following parental death; elder siblings sometimes asserted an authoritative parental role to provide moral guidance and ensure that younger siblings respected them, helped with household chores and did not misbehave, while at other times, older and younger siblings advised each other and discussed and planned their lives together on a more equal basis. Such sibling practices reveal a concern to continually re-make and strengthen family relations to ensure younger siblings' moral upbringing and the survival of the group, which was critical in child- and youth-headed households, within a wider context in which such households were constructed as deviant. Indeed, some young people heading households with little access to external support felt that had they not cared for their siblings, they might have already died, due to their young age and the lack of adult relatives or neighbours who were willing or able to care for them.

The findings point to the importance that young people placed on expressing connections and continuities with the care provided by their deceased parents, evidenced in their concerns to fulfil their deceased parents' wishes to care for siblings, to manage the household in the same way that their parent had and to safeguard their inherited parental home and land. The findings thus provide a fresh perspective on how to understand the materiality of care after death and 'continuing bonds' with the deceased in the context of the global South. Families often had few material belongings and assets to pass on to orphaned youth and little money for elaborate graves or other memorials, which in more affluent countries may represent an important means of expressing continuing bonds and memorialising the deceased (Maddrell 2013). In such contexts, staying together in the inherited parental home, maintaining parents' (simple) graves and cultivating inherited agricultural land were important ways for young people to express 'continuing bonds' with the deceased. Young people's efforts to demonstrate the values and behaviour they had inherited from their parent can also be regarded as an expression of 'continuing bonds' and a means of 'displaying' family and the legacy of their deceased parent to each other and to those outside the household.

This chapter has thus revealed that analysing intra-generational relations following parental death and young people's position within hierarchies of sibling birth order, age and generation offers a rich understanding of how 'continuing bonds' and care, as well as families, are 'done' and 'displayed' in diverse spatial and temporal contexts. While childhood studies have demonstrated young people's active roles in caring for parents and other adult relatives with a disability or chronic illness, this chapter calls for more attention to be paid specifically to *sibling practices* and *caregiving* in both the global North and South. There is also much scope for death and bereavement research to pay more attention to everyday family contexts of responses to death, caring relations and continuing bonds in diverse cultural contexts (Evans et al 2016). Sibling practices performed in varying household spaces and circumstances, including following the death of a relative, reveal much about the complexity of care, inter- and intra-generational relations and socio-cultural meanings of 'family'.

Acknowledgements

I would like to thank all the young people, relatives, community members and professionals who participated in the East African and Senegal studies reported here. I am grateful to the Royal Geographical Society (with the Institute of British Geographers), the University of Reading and the Walker Institute for funding the two studies. I am also grateful for research assistance and facilitation from Germina Sebuwufu, Mbelwa Gabagambi, Judith Nakayiza Mukasa, Baba Sarr,

Fatou Ke´be´, Prosper Diene Faye, Khady Sarr, Soguy Ndiaye and Yacine Diagne. I also thank Jane Ribbens McCarthy and Sophie Bowlby for our discussions as part of our research project, *Death in the Family in Urban Senegal: Bereavement, Care and Family Relations* (Evans et al 2016), which have helped to inform this chapter. Figure 10.1 has previously appeared in Evans, R. (2011) 'We are managing our own lives…': life transitions and care in sibling-headed households affected by AIDS in Tanzania and Uganda. *Area* 43, pp.384–96. We thank John Wiley and Sons for permission to republish this image under license number 3847671247604 (April 2016).

References

Abebe, T. (2012) AIDS-affected children, family collectives and the social dynamics of care in Ethiopia. *Geoforum* 43, pp.540–50.

Barnes, M. (2012) An ethic of care and sibling care in older age. *Families, Relationships and Societies* 1, pp.7–23.

Bauman, L., Foster, G., Johnson Silver, E., Gamble, I. and Muchaneta, L. (2006) Children caring for their ill parents with HIV/AIDS. *Vulnerable Children and Youth Studies* 1, pp.1–14.

Brannen, J., Heptinstall, E. and Bhopal, K. (2000) *Connecting children: children's views of family life*. London, Falmer Press.

Burke, P. (2004) *Brothers and sisters of disabled children*. London, Jessica Kingsley.

Carsten, J. (2004) *After kinship*. Cambridge, Cambridge University Press.

Cicirelli, V. (1994) Sibling relationships in cross-cultural perspective. *Journal of Marriage and the Family* 56, pp.7–20.

Cluver, L., Gardner, F. and Operario, D. (2007) Psychological distress among AIDS-orphaned children in urban South Africa. *Journal of Child Psychology and Psychiatry* 48, pp.755–63.

Collard, D. (2000) Generational transfers and the generational bargain. *Journal of International Development* 12, pp.453–62.

Dahlblom, K. (2008) *Home alone. Sibling caretakers in Léon*. Nicaragua. Unpublished dissertation, Department of Public Health and Clinical Medicine, Umeå University, Sweden.

Demmer, C. (2007) Grief is a luxury: AIDS-related loss in South Africa. *Illness, Crisis and Loss* 15, pp.39–51.

Edwards, R., Hadfield, L., Lucey, H. and Mauthner, M. (2006) *Sibling identity and relationships: sisters and brothers*. Abingdon, Routledge.

Evans, R. (2004) Tanzanian childhoods: street children's narratives of 'home'. *Journal of Contemporary African Studies*, 22 (1). pp. 69–92.

Evans, R. (2010) Children's caring roles and responsibilities within the family in Africa. *Geography Compass* 4, pp.1477–96.

Evans, R. (2011) 'We are managing our own lives…': life transitions and care in sibling-headed households affected by AIDS in Tanzania and Uganda. *Area* 43, pp.384–96.

Evans, R. (2012a) Sibling caringscapes: time-space practices of caring within youth-headed households in Tanzania and Uganda. *Geoforum* 43, pp.824–35.

Evans, R. (2012b) Safeguarding inherited assets and enhancing the resilience of young people living in child- and youth-headed households in Tanzania and Uganda. *African Journal of AIDS Research* 11, pp.177–89.

Evans, R. (2014) Parental death as a vital conjuncture? Intergenerational care and responsibility following bereavement in Senegal. *Social and Cultural Geography* 15, pp.547–70.

Evans, R. (2015a) Negotiating intergenerational relations and care in diverse African contexts. In Vanderbeck, R. and Worth, N. (eds.) *Intergenerational space*. London, Routledge, pp.199–213.

Evans, R. (2015b) HIV-related stigma, inheritance and chronic poverty: the vulnerability and resilience of widows and caregiving children and youth in Tanzania and Uganda. *Progress in Development Studies* 15, pp.326–42

Evans, R. (2016) Gendered struggles over land? Shifting inheritance practices among the Serer in rural Senegal. *Gender, Place and Culture*, online early.

Evans, R. and Thomas, F. (2009) Emotional interactions and an ethic of care: caring relations in families affected by HIV and AIDS. *Emotions, Space and Society* 2, pp.111–9.

Evans, R., Ribbens McCarthy, J., Bowlby, S., Wouango, J. and Kébé, F. (2016) *Responses to death, care and family relations in urban Senegal*, Draft Research Report 1, Human Geography Research Cluster, University of Reading.

Finch, J. (2007) Displaying families. *Sociology* 41, pp.65–81.

Francis-Chizororo, M. (2008) *The formation, constitution and social dynamics of orphaned child-headed households in rural Zimbabwe in the era of HIV/AIDS pandemic*, unpublished PhD thesis, School of Geography and Geosciences University of St. Andrews, Scotland.

Hall, S.M. (2016) Moral geographies of family: articulating, forming and transmitting moralities in everyday life. *Social and Cultural Geography*, online early.

Hockey, J., Komovomy, C. and Woodthorpe, K. (2010) *The matter of death: space, place and materiality*. Basingstoke, Palgrave MacMillan.

Howarth, G. (2007) The rebirth of death: continuing relationships with the dead. In Mitchell, M. (ed.) *Remember me: constructing immortality. Beliefs on immortality, life and death*. London, Routledge, pp.19–34.

Kesby, M., Gwanzura-Ottemoller, F. and Chizororo, M. (2006) Theorising other, 'other childhoods': issues emerging from work on HIV in urban and rural Zimbabwe. *Children's Geographies* 4, pp.185–202.

Klass, D., Silverman, P. and Nickman (eds.) (1996) *Continuing bonds. New understandings of grief*. New York, Taylor and Francis.

Klett-Davies, M. (ed.) (2008) *Putting sibling relationships on the map: a multi-disciplinary perspective*. London, Family and Parenting Institute.

Kosonen, M. (1996) Siblings as providers of support and care during middle childhood: children's perceptions. *Children and Society* 10, pp.267–79.

LeVine, R., Dixon, S., LeVine, S., Richman, A., Leiderman, P., Keefer, C. and Brazelton, T. (1996) *Child care and culture: lessons from Africa*. Cambridge, Cambridge University Press.

Maddrell, A. (2013) Living with the deceased: absence, presence and absence-presence. *Cultural Geographies* 20, pp.501–22.

Maddrell, A. and Sidaway, J. (2010) *Deathscapes: spaces for death, dying, mourning and remembrance*. Farnham, Ashgate.

Mauthner, M. (2005) Distant lives, still voices: sistering in family sociology. *Sociology* 39, pp.623–42.

Meintjes, H., Hall, K., Marera, D. and Boulle, A. (2010) Orphans of the AIDS epidemic? The extent, nature and circumstances of child-headed households in South Africa. *AIDS Care* 22, pp.40–9.

Morgan, D. (1996) *Family connections: an introduction to family studies*. Cambridge, Polity.

Morgan, D. (2011) *Rethinking family practices*. Basingstoke, Palgrave MacMillan.

Nyambedha, E., Wandibba, S. and Aagaard-Hansen, J. (2003) Changing patterns of orphan care due to the HIV epidemic in Western Kenya. *Social Science and Medicine* 57, pp.301–11.

Omari, C. and Mbilinyi, S. (1997) *African values and child rights: some cases from Tanzania*. Dar es Salaam University Press, Dar es Salaam.

Payne, R. (2009) *Child-headed households in Zambia: from concepts to realities*. Unpublished PhD thesis, Department of Geography, Royal Holloway University of London.
Punch, S. (2001) Household division of labour: generation, gender, age, birth order and sibling composition. *Work, Employment and Society* 15, pp.803–23.
Punch, S. (2008) Negotiating the birth order: children's experiences. In Klett-Davies, M. (ed.) (2008) *Putting sibling relationships on the map: a multi-disciplinary perspective*. London, Family and Parenting Institute, pp.30–49.
Rabain-Jamin, J., Maynard, A. and Greenfield, P. (2003) Implications of sibling care-giving for sibling relations and teaching interactions in two cultures. *Ethos* 31, pp.204–31.
Ribbens McCarthy, J. (2009) Young people making meaning in response to death and bereavement. In Balk, D. and Corr, C. (eds.) *Adolescent encounters with death, bereavement and coping*. New York, Springer, pp.21–38.
Ribbens McCarthy, J. (2012) Caring after death: issues of embodiment and relationality. In Rogers, C. and Weller, S. (eds.) *Critical approaches to care*. London: Routledge, pp.183–94.
Skovdal, M., Ogutu, V., Aoro, C. and Campbell, C. (2009) Young carers as social actors: coping strategies of children caring for ailing or ageing guardians in Western Kenya. *Social Science and Medicine* 69, pp.587–95.
Stalker, K. and Connors, C. (2004) Children's perceptions of their disabled siblings: 'she's different but it's normal for us'. *Children and Society* 18, pp.218–30.
Thompson, R., Bell, R., Holland, J., Henderson, S., McGrellis, S. and Sharpe, S. (2002) Critical moments: choice, chance and opportunity in young people's narratives of transition. *Sociology* 36, pp.335–54.
Thurman, T., Snider, L., Boris, N., Kalisa, E., Mugarira, E., Ntaganira, J. and Brown, L. (2006) Psychosocial support and marginalization of youth-headed households in Rwanda. *AIDS Care* 18, pp.220–9.
Vanderbeck, R. and Worth, N. (eds.) (2015) *Intergenerational space*. London, Routledge.
Weisner, T. (1982) Sibling interdependence and child caretaking: a cross-cultural view. In Lamb, M. and Sutton-Smith, B. (eds.) *Sibling relationships: their nature and significance across the lifespan*. London, Lawrence Erlbaum, pp.305–27.

11 Kindness

Caring for self, others and nature – who *cares* and *why*?

Tracy Hayes

Introduction

Kindness is a behaviour defined by ethical characteristics including a pleasant disposition, care and concern for others, and is demonstrated by acting in a friendly, helpful, considerate manner (OED 2011). In some cultures, it is known as a virtue, and in others as a value; I see it as a moral stance. We can be kind to ourselves, to others and towards nature. I will now be kind to you, the reader, by explaining my approach to this chapter, which adopts a friendly, narrative style to explore caring for self, others and nature. It is informed by both social and natural science and presents a philosophical, conceptual and empirical discussion. Like many concepts, for example sense of place, care is defined, conceptualised, theorised and applied "across different disciplines in different ways" (Convery et al 2012, p.1). Consequently, I have adopted an interdisciplinary approach, which I will explain in more detail later. My language is deliberately evocative, provocative and multifaceted; it is "scholarship [that] reaches toward an audience, it cares. It wants to make a difference" (Pelias 2004, p.12). It is written in response to calls for more "diverse, challenging, exciting, creative, and interdisciplinary work" (Kraftl et al 2014). Like many researchers exploring the slippery stuff of the social world (for example, Macartney 2007, Wood 2007, Jones 2000, 2008) I often feel constrained within traditional forms of both qualitative and quantitative methodologies. Consequently, this chapter takes a boundary-crossing (Ellis 2004) integrated, yet fully-contextual approach to both empirical research and to literature, drawing from the wells of geography, environment and nature; youthwork; experiential, outdoor and adventure education; child/adolescent development and play; storytelling and creative writing; philosophy and psychology.

The chapter begins with a fable, a short story that serves a moral purpose, inviting moral contemplation, but not judgement; the fable is based on my empirical research with 'looked after' young people. This fable is then utilised in stages to explore the concept of care, before finally being placed within a wider context that invites the reader to also question what we mean when we use the word 'care'. The story is one of many 'magic moments' that I have experienced as I have explored young people's relationship with nature. I am exploring how young people make sense of outdoor experiences that have been provided for them as part of a targeted, specifically funded intervention. I am curious as to how these experiences impact on their perception of their own health and wellbeing, and on their desire to look after (to care-for) outdoor spaces, which in everyday terms we call 'nature'. My research is interdisciplinary (or transdisciplinary), in that I have made use of an integrated blend of disciplines from the outset (Kessel et al 2009); this is not surprising in view of my earlier training, first as a natural scientist, then as a social scientist. This particular piece of research has arisen from my previous practice as an environmental youthworker, and aims to discover what it is that we need to do to make sure our work is effective – so that young people and practitioners find value and meaning in their experiences, and there is an increased awareness of the need to conserve our natural environment.

My research has involved spending time with young people and practitioners participating in a number of different projects, across a range of settings: urban, rural and coastal, from the south-east to the north-west of England. All of the young people were aged between 11 and 21, and had been identified as having specific needs that would benefit from a targeted intervention, with pre-determined outcomes. This included young people with special educational needs / disabilities (SEN/D); young people classed as NEET (not in employment, education or training); young people following an alternative curriculum, both within statutory education (schools and colleges) and in alternative, community-based education; young people attending community-based youth clubs, and visiting residential outdoor centres; young parents; young people in care/preparing to leave statutory care. As I write this chapter, I reflect on the young people who have participated in the projects that have been included in my research; although they are far from a homogenous group, indeed, they are a wonderfully diverse and eclectic group of young people, what they share is the experience of exclusion, of being marginalised from the mainstream world. As a result of this, many of them have experienced considerable unkindness and a lack of care: I have no wish to inadvertently repeat that in the way that I interpret and write about my research, which again reiterates the need to take an interdisciplinary approach to gain a more holistic understanding.

My findings have highlighted a need for playful, creative and emotional approaches that allow young people to feel comfortable *being* in nature, rather than always *doing* things to, in and for nature. The natural world can be perceived to be a magical and enchanting place for children; indeed, many of the activities we offer as practitioners actively encourage this as they are based on imaginative play. However, as children grow older and develop an awareness of the state of the natural world, they begin to understand the urgency of the need to protect it, and as a consequence, the 'magic' often disappears to be replaced by more serious considerations and concern. In my approach to addressing this, I have been inspired by writers such as Rachel Carson (1962/2000) to create fables designed to stimulate interdisciplinary dialogue in an engaging and understandable way. This has also necessitated challenging more traditional forms of research and writing, by embracing more 'stylish' approaches (Sword 2012), primarily by making my work accessible through avoiding technical jargon or overly academic language, and maintaining my focus on practical applications and everyday language.

The moments used for my fables are ones that I have perceived as epiphanies, turning points or an address (Moules et al 2014). They have proved to be the defining moments that have made me stop, wonder and question in an attempt to extract meaning (Denzin 2014). They are the times when I have felt bothered by something, have become 'troubled' (Meyer and Land 2003). When this happens I take care to critically analyse each specific moment, to carefully avoid it being merely 'a nice story'; it is so much more than that. It is an opportunity to "illustrate facets of cultural experience … make characteristics of a culture familiar for insiders and outsiders" (Ellis et al 2011, p.3). Like most of my magic moments, there was nothing momentous, or out-of-the-ordinary about this particular experience, indeed it was arguably, its very ordinariness (Ross et al 2009), its routineness (Denzin and Lincoln 2011), its everydayness (Horton and Kraftl 2006) that made it so special. It occurred in the early stages of my empirical research, and it is one that has sustained and nurtured me throughout. I offer it to you here and now in story form and invite you to slow down your mind and let my words sink beneath your skin (Hustedde and King 2002) so that you may also reflect on the meanings of this story, and consider how to relate this approach to exploring your own experiences. This particular story is dedicated to Liz, the pseudonym chosen by the young man at the centre of the story, and utilised to address moral and ethical dimensions of empirical research. I will tell you more about him (and that) later, but for now, let me share his (our) story.

> "*But Tracy, I don't get it. I hear Londoners who come here on holiday, I see them looking at everything, I hear them say 'Wow! Look at that mountain, that lake, wow!' But I just don't get it. I don't see it - I think it's cos I grew up here, it's just normal, nothing special. I don't say wow …*"
>
> His voice tails off as his thoughts drift to another place. "*I do,*" I reply, "*I live here and I still say wow – nature is so wonderful. What do you think it would take for you to say wow*?" Looking at me sadly, shaking his head, he responds, "*Nothing, I just don't get it. I don't feel the 'wow thing'. It's just outside. I do that all the time, I like being inside …*"
>
> Our conversation moves on through a discussion of some of the things planned for this residential weekend: a combination of two types of activities. First, being trained by me to be a peer mentor to support other young people who are in statutory care, preparing to leave or who have just made the move to independent living. Secondly, to participate in a number of adventurous outdoor activities, facilitated by colleagues. Discussion widens to include the other young people sitting in the back of the mini-bus alongside us, with topics ranging across music, film and sport. We arrive at the centre and settle into our designated spaces: separate rooms for boys and girls, with leaders in between. Dinner is served, eaten and the table cleared, then we head out for an evening walk. Each participant has been equipped with walking boots,

> waterproofs and head-torches. It has been raining and the ground is sodden and heavy underfoot. Feet in badly fitting boots slip, trip and stumble, accompanied by squeals and laughter. Two of the young men offer me their arms, jokily giving support to the 'old mother' of the group …
>
> We stop part way up the fell. Each person is placed by my colleague in their own special spot, to experience the quiet of the fell on their own. The more nervous ones are reassured that we are nearby. Then we wait. After a while, ten minutes or so, we gather them up and make our way back down the fell, to a space where we can reflect on the activity and share our experiences. And then it happens, the word escapes his lips "*Wow! What's that*?" He is pointing to the top of the fell, to where a gentle light is appearing. "*That's the moon*," my colleague explains, "*The moon is rising behind the fell and casting its light across the top. If we wait a bit, we'll see the moon*." We wait and watch, then again I hear it, softly as if forming part of an involuntary breath: "*wowwwww*". Quietly, we return to the centre. Talking later about the experience I gently ask him, "*Do you know what you said, up there on the fell*?"
>
> He doesn't, he is unaware …

The ellipsis points at the end of four sentences within this story signify places at which to pause and focus the conversation; this approach conceptualises place as somewhere to stop, rest and become involved. There are other places I could have chosen (for example, home making/breaking; silence; gender/sexuality) and I encourage you to consider which you would have chosen if I had not made this choice for you, if I had left you alone to interpret the story in your own way. Instead, I facilitate the exploratory process by asking specific questions, a process which culminates in addressing the story's final thought. In doing this, I am embracing the trend for "conceptual approaches to closely align with the nature of changing bodies of knowledge in a fluid social world … conceptual understandings are better understood as transition points rather than endpoints" (Milligan and Wood 2009, p.2). In the hermeneutic tradition of 'questionableness' (Nixon 2014), the chapter will end, not with conclusions, but with more questions to encourage further contemplation. I make no apology for this: the world is full of messiness (Huestedde and King 2002; Jones 2008) and does not fit within 'neat' conclusions; I believe it is morally wrong to pretend otherwise.

A brief introduction to Liz before we explore the story: at our time of first meeting, he was 17 years old and had been in statutory care since the age of six. We met through my role as a researcher responsible for evaluating a Cumbria-wide service for young people who were identified as being in care, or having just left care. The service had the espoused aim of easing their transition to independent adult life, occurring during the phase of life that sees a shift in focus away from being in care and towards caring for self. This transition is something acknowledged as being of international concern, resulting in disrupted education and poor social outcomes (Rogers 2011). He informs me that he has several siblings, most of which he does not see, or have contact with. I do not pry, or seek background information from colleagues; as a professional youth worker I prefer to get to know him on his terms, learning about him as and when he is ready. However, I have met one sister, who also engages with the project where I am based. I do not know her so well; she is more reluctant than him to join in with activities. He is a gentle person, openly conflicted about his sexuality, and very unsure about his future; he is conscious that his time in care is coming to an end, and that he is moving towards independence and a less supervised kind of life. Unlike many 17-year-olds who look forward to their 18th birthday with great anticipation, he views it with trepidation as it will mean that his visits with his mother will no longer be supervised. He shares that he is not ready to face this alone; action by his social worker ensures that this is delayed until he is ready. I enjoy his company; he has a well-rounded sense of humour, and a seemingly never-ending bank of funny train stories to share with me: things he has seen or heard or experienced whilst travelling. When he found out about my research project, he was determined to be included and was one of the first young people I interviewed; part of that research conversation is presented within this story. In the mini-bus, on the way to the residential, we had talked about what I was doing, with him asking me lots of questions about the why (purpose) and how (methodology), which had prompted his first remarks.

I don't say wow …

Wow is a universally recognised word uttered as an exclamation of astonishment, admiration, awe, wonder or joy. Liz, in the moment, on the bus, does not feel any of this, more a sense of bewilderment, and I perceive a feeling of loss, of sadness. This small three-letter word represents the effect of these emotions; as explained by Fuller (2006) they excite our imaginations, enhance "our capacity to seek deeper patterns in the universe" (ibid, p.2) and provoke reflection and contemplation. When we encounter something that we perceive to be new or very different, it surprises us, it causes us to wonder. Descartes identified that this happens before we know whether this object/thing will be of any use to us: it is a spontaneous and intuitive response to something different, something wonderful. Focusing on wonder as a way of exploring this, Fuller explains that wonder is a form of heightened interest, which "momentarily suspends habitual ways of looking at the world and instead lures people into new and creative engagement with their surroundings … wonder induces receptivity and openness. It prompts us to become more connected with the wider environment" (Fuller 2006, p.12). Liz currently lives in what most people consider to be one of the most beautiful parts of the UK, and yet he does not appear to notice or appreciate this. It is generally accepted that early experiences in nature are important for developing an awareness of, and a connection to nature (see for example Chawla 1990, Waite 2007, Dunlap and Kellert 2012). This is nothing new: in 1956, Rachel Carson encouraged parents and carers to take children outside, highlighting that "a child's world is fresh and new and beautiful … full of wonder and excitement […] Daily, intimate, sensory experience is essential to keep that wonder bright for a lifetime" (Carson 1956, pp.42–3). The most effective way for this to happen is through companionship with a responsive adult, however what happens to the young people who enter adulthood never having experienced nature in this way? Who helps them to discover the wonder and then keep it bright (Carson, *op. cit.*)? I have written elsewhere about the importance of playfulness in addressing this (Hayes 2013, 2015) however it takes more than a playful approach: it takes kindness and care.

Retracing our steps to the start of this section of the story, Liz's choice of wording to describe those who do say wow as 'Londoners' is interesting. In this one term he encapsulates an extreme sense of 'otherness': London is a very long way from the west coast of Cumbria where Liz spends much of his time. Applying a number of geographical concepts (see suggested further reading), as a place it is spatially, temporally and culturally distant and different, and the people who come from there are different to Liz: there are "vast distances in knowledges, needs, modes of being and experiences" (Jones 2008, p.200). If we choose to apply socio-cultural concepts, we could consider the added dimension of how 'Londoners' are perceived as different: more affluent, mobile, professional – visitors coming to gaze in wonder at the landscape, or second-home owners who make living unaffordable for many or adventurous types keen on ticking off yet another mountainous peak. However, those are complex topics and issues to savour and explore elsewhere. Here it is sufficient to say that Liz sees them as 'other', as unrepresentative of him and his life: they are distinctly different.

Unlike Liz, I did not grow up in Cumbria; in my early adulthood I lived and worked near London, and two of my grown-up children currently live there. Cumbria is a place where I have chosen to live, moving to make a permanent home after many years of holidaying here. I feel a kinship with the 'Londoners' who say 'wow': at the time of the encounter I had been resident here for less than a year, and even now I am still regularly surprised by what I see around me. Liz has not yet had the chance to choose his home: he started life within a family home which was abruptly disrupted and disturbed when the children were removed from the care of their parents. Separated from parents, and siblings, he was placed elsewhere, in homes and with people that were chosen for him. Within a later conversation he confides, "*I remember being taken into care. No one told me anything except, 'you have three minutes to pack*'". Conscious of how long it took me to carefully pack the numerous boxes that I filled with my treasured belongings (particularly books, teddy bears and toys from my childhood, and my children's), and what I paid to have them safely transported north to Cumbria to enable me to meaningfully set up my new home, I am shocked at how little he would have been able to take with him. Three minutes? This was all the time he was given to gather his belongings before being taken away in a police car to a 'family' he did not know. I am deeply troubled by this, and find myself wondering what would I grab if that was all the time I was given? Not having lived his experience, I am unable to find a specific answer to this question, yet my emotional response allows me to imagine what it may have been like and to empathise. Responding to his words, his story, I feel a sense of loss and sadness – interwoven with gratitude for having the luxury of time to pack. I share this with you to further your understanding of the context for this story and to enable you to test out your personal response to it (Smith and Smith 2008).

Liz was given a social worker, although he did not know what one was, who informed him "You will now be in care until you're 18 years old". That did not mean anything to him either; he was too young to understand. This marked the beginning of his life in care: being placed in a new house, away from people he knew, with few belongings, little comfort or sense of belonging. No wonder he does not say "*wow*" when he steps outside. He does not see the inauthentic, romanticised, rural idyll (see, for example Jones 2000, 2008, Valentine 2001) that we (as in me and the 'Londoners') do. His view is very different to ours; whilst we are gazing in awe at a landscape that has been commodified for our pleasure and leisure (Valentine 2001), he is looking at a much rawer, gritty and disturbing space, where bad, unkind things happen to children like him.

It's just outside. I do that all the time, I like being inside …

Outside/inside and insider/outsider are terms that invite us to explore through metaphorical, philosophical and geographical concepts (see for example, Bachelard 1958/1994). Here we are going to accept the everyday usage of Liz. Like many people (young and old) Liz spends much of his time indoors, either in his foster home, on the train or at college. He also volunteers at a local charity shop; you could say that his life is full of indoor learning. There are numerous studies, theoretical and empirical, that highlight our concerns about how little time westernised children and young people spend outside (see for example Louv 2005, White and Stoecklin 2008, Ridgers et al 2012). As explained in detail by Palmer (2006), this is one element of what is seen by many as an alarming deterioration in children/young people's understanding and appreciation of the world they live in. In her book she urges us (people who facilitate learning, whatever our defined role may be) to enable children and young people to play outside more. This viewpoint is substantiated by research; for example, in 2012 Sigman reported some alarming statistics which highlighted that by the age of seven, the average westernised child will have spent the equivalent of one whole year of 24-hour days watching screen media; by 18, Liz's age, this increases to the equivalent of three years' worth (Sigman 2012, pp.8–9). Having spent time with him, he does not appear to conform to this; he has an old style phone, which, unlike others within the group, tends to stay out of sight (perhaps he may be embarrassed by it?), however he demonstrably enjoys spending time with people, engaging in conversation and eye-contact. He actively seeks out opportunities to participate in activities – both indoors and outdoors.

In the mini-bus he was quick to inform me that he liked being indoors, but in a later reflective conversation, he contradicts this, telling me that young people "*think they're a lot safer indoors, in their own surroundings. But actually you're a lot safer outside*". Is this a result of the training we have provided as part of the peer-mentoring programme? He has learned about the need for risk assessments, professional boundaries, and appropriate behaviour. He has a greater awareness of what we, as facilitators of the experiences, actually do and an insight into why we do this the way we do. This appears to have altered his perception – although perhaps it may be because he knows me better, feels able to share more sensitive memories and thoughts. As an adult I am unable to confidently explain this, so I choose to accept it as it is, with all its messiness in relation to my empirical research. In this moment of speaking with me, Liz appears to feel safe, which is what matters most to me.

Jokily offering support to the 'old mother' of the group …

In this moment, I have to ask myself exactly who is supporting whom? Looking back on the experience as I write this chapter, I remember the feeling of wet mud, clinging to my boots, sucking them under and causing me to wobble; the firm grasp of hands, appearing on either side gripping my arms to keep me upright; the shrieks as one of my helping hands (standing to the left of me) slips in his badly fitting, borrowed boots, nearly taking me with him. This is quickly followed by laughter and reassuring words, "*You're safe Tracy, I won't let you fall*". I recall looking up at the earnest young face to my right, it is Liz, and I recognise that his slight build is highly unlikely to hold me (and my much more considerable weight), should I fully lose my footing. We pause, then linking arms, three of us slowly move forward together, more stabilised than when we were walking alone. Ahead we can just about make out the shapes of others in the group: the experienced outdoor leader and other young people. They do not appear to be struggling as much as us, striding confidently out across the muddy field. Our physical closeness enables us to talk and our conversation is relaxed and humorous; we tell each other silly childish jokes and they refer to me as the old mother of the group. The following day, on their return from the activities they have undertaken with my colleague (I stayed behind to mark their coursework), they respond to me greeting them at the door with the remark, "*Oh, look it's Old Mother Hubbard*". I laugh as I move to switch on the kettle to make them a drink. Even now, several years on from the experience, I remember the warmth – both emotional and physical – between us and I realise that I miss their company; we have all moved on in very different ways. I wonder what they remember from the experience.

Rousing from my reveries, and returning to the story, an unintended consequence from losing my footing was the opportunity for them to show how much they cared about me, and in this instance, they cared-for me: we were moving towards being in a reciprocal, caring relation. Like Pyer (2008), in her study with young people who are wheelchair users, I am conscious of the potential we have as researchers to affect others. Is it fair to allow a caring relation to develop, when our time together is time-limited, particularly for those who lives have involved many transient relationships? Would it be kinder (for them and for me) not to show this type of care? I agree with Pyer (2009, p.215), with regards to duty of care "sometimes there are legacies that ethical researchers must endeavour to deal with", which I am attempting to do through critical and theoretical reflections to identify the learning to take forward. Also, it is only with the benefit of hindsight that I realise the importance of humour and laughter within these experiences; like Macpherson (2008), this was not something I set out to explore, and yet the more I read and reflect on my 'magic moments', the more I listen and attend to voice recordings and transcripts of interviews, the more conscious I become of the *funniness* involved.

Considering my role, I am very adept at holding coats, taking photographs, and providing support for those who wish to be adventurous; in many ways, I would describe myself as adventurous, certainly in terms of my research and teaching, however this does not extend to risky physically strenuous activities. I prefer adventures that are slower, more gentle and inclusive (Hayes 2014) usually accompanied by my teddy bear (Hayes 2015) or bunny rabbit (Hayes 2013)! On this particular residential experience, I was happy to play the supporting role, although also highly conscious that the aim of the project was to encourage independence. I felt a tension between two apparently binary positions: wanting to *care-for* these young people, who have experienced a lack of care in their lives; and caring *about* them, in a less involved, more professionally detached way, that acknowledged the restrictions inherent in our roles. Noddings (2013) clearly elucidates this distinction by describing caring-for as "an encounter or set of encounters characterized by direct attention and response. It requires the establishment of a caring-relation, person-to-person contact" (ibid., p.xiv). She asserts that it is impossible to care-for everyone; we do not have sufficient time or resources for this. The awareness of this distinction can help us to be more effective in our work. My colleagues are adept at caring about people, and they take great care to organise activities so that risk is managed in a way that allows for risky activities to be enjoyed; their role does not allow them to care-for the participants, they adopt a less emotional stance. I take care to avoid these activities; and yet I embrace the equally risky world of emotions, and prefer situations where I am able to care-for people. Self-awareness, including why we choose to do certain things and make particular career choices, is integral to working effectively.

As highlighted by Marschall (2014, p.518) within a study of children's experiences following the divorce of their parents, the dominant discourses around care have developed from the normative perspective of the nuclear family, and "draw on general and rather implicit assumptions of caring as a unidirectional, linear and causal process of caregiving from parents". As we can see from the story in this chapter, and from Marschall's own research, this is not really the case: care is not a linear, one-way process, it is relational, situational and complex; it includes intrinsic elements of human development such as attention, love and kindness, as well as care. Attempting to address these needs is therefore not easy; it is so much more than just 'give and take'.

He doesn't … He is unaware

Despite his earlier protestations, Liz says the word 'wow' twice. The first time in surprise and curiosity, directing our attention to what he has seen. The second as an almost involuntary breath, an instinctive, intuitive response to what he has encountered. It is nothing particularly exceptional or unexpected, however it is the first time that Liz has noticed it, has paid attention to it and wants to share it with us. He was transfixed by the moonlight – he could not see the moon as it had not risen high enough, however it cast its light and its shadow over the fells. He had spoken of wanting to stay on the fell and watch it until morning. But the clouds descended and the moon disappeared, and the moment passed into memory. As identified by Sissay (2012), in general the "… great moments of childhood are recalled through family members"; my efforts to capture it within a fable can be seen as an attempt to make up for this absence. If we consider the role of the practitioner, as the facilitator, co-participant and interpreter of experiences, we have a privileged 'insider' role that allows us to share these magic moments. I suggest we have a duty of care to ensure that we do this.

And yet as we recall those moments of childhood, we are looking back through time, through the eyes and awareness of an adult. As identified by Jones (2003, p.34) when doing this "our imagination needs to work with our memory and we need to recall the feelings and emotions themselves, as far as is possible, as well as narrative accounts of events". Does it matter if the memories that Liz shares with me may not be entirely historically accurate? Does anyone care that his account may be contradicted by other family members, or people in positions of authority responsible for his care? Is this an important dimension of empirical research? Not to me, because I am interested in his perception of his experiences as shared by him.

Like health professionals (see for example, Bray et al 2013) practitioners of outdoor experiences need the ability to show kindness and compassion through caring-for, in addition to more professional skills and knowledge. In Bray et al's study into the role of professional education in developing compassionate health practitioners, they identify compassion as a core value which represents "responding with humanity and kindness to each person's pain, distress, anxiety or need […] respecting and valuing people as individuals and responding to them in ways that appreciate the human experience" (ibid., p.481). Although their study focused on what we recognise as the health service (i.e. nursing) their findings are applicable to our more general wellbeing and to the role of other practitioners. Kindness encapsulates the relational reality of caring: its reciprocity, responsiveness, attentiveness and care.

Care … who cares, about what, why, when, where and how …

Retracing my steps once more to the title of the article, so many questions may be generated by one very small, taken-for-granted word: care is an everyday word, with an everyday meaning – or is it? Like much in life it all depends on how it is applied (OED 2011): as a noun it implies serious attention, thought, caution, protection, worry or anxiety; as a verb it signifies concern, interest, affection, liking or willingness – for example, would you care to accompany me as we explore this word further? We can join it to other words, and become careless, careful, carefree, carer, caregiver, caretaker, careworn; we can care for, be cared for, care about, be cared about, come-in-to-care, be in care, prepare to leave care and become a care-leaver. This last role is a lifelong one: as highlighted by Lemn Sissay (2013, unpaginated) who states "I am a care-leaver for the rest of my life. It's not something I get over or avoid. It's who I am." This one aspect, which is not of our own making, becomes embedded within our identity, a socially constructed part of who we perceive ourselves to be. At this point I am reminded by the words of Lorde (1984, pp.120–1), to take great care: when we are forced to focus on one part of ourselves, categorised (labelled?) by an 'externally imposed definition', this one part of our identity may be presented as "the meaningful whole, eclipsing or denying the other parts of self … this is a destructive and fragmenting way to live". We are so much more than that one part. We need to look beyond the traditional, hegemonically-derived roles of care-giver (parent/carer/practitioner) and cared-for (child/young person) by emphasising that this is a reciprocal process, involving practices that are actively negotiated and adjusted by both sides, across space and time. The balance may shift from 'side to side' depending on the specific situation and context, but each 'side' is of equal value in this process of caring, of recognising and attending to each other's specific needs, demonstrating kindness. This is the relational reality, it is "inherently dynamic and interrelated at all levels" (Spretnak 2011, p. 1).

To be *in care* means that you are no longer cared for within your own family, or by a family member, or even in your own home; you are in the care of your local authority, in statutory care, in that space that someone else calls home, and tells you to do the same when you are placed there. This is very different. In the UK, it has been reported (see for example, Rogers 2011) that the main reasons for children coming into care are abuse/neglect (60 per cent) and 'significant family dysfunction', typically related to alcohol and/or drug abuse (10 per cent). If we accept Noddings' argument that we learn to care through watching and modelling the caring practices of others, combined with opportunities to "have supervised practice in caring" (2013, p. xviii) then it becomes even more important to consider our own practices. By allowing them to care-for us, we are supporting their development; this is not a sign of weakness on our part, but a very important lesson for life and part of a reciprocal caring relation. In Rogers' study, young people "described how they had felt isolated, abandoned and in some instances betrayed by their care givers during and following the period in which they became care leavers" (Rogers 2011, p.417). Her research concluded with a powerful statement that what young people most associated within the challenges of leaving care was the "removal of emotional support (more than practical support or advice)" (ibid., p.423). It is vital therefore to take full account of the importance of emotions, and look to provide "enabling research environments, encounters and exchanges" (Ross et al 2009, p.605). In this way we can prioritise the development of caring relations, so that young people are both cared-for and cared-about, and are enabled to care-for themselves and others.

Returning once more to the story to explore the 'otherness' of those who say 'wow', I am approaching this word in the footsteps of Jones (2008, p.196), seeing it as a way of sidestepping the "adult agendas which seek to colonise and control childhood" by favouring an approach that acknowledges "the background field of ethical, methodological and ontological fabric of any academic endeavour". The 'Londoners' of which he speaks are adults (and their accompanying children) who have come to 'look', to walk, and are symbolic of those identified by Haywood (2012, pp.23–4) as having created the 'Lake District': each "circular journey undertaken on foot has become an important constituent of the Lakes' transformation from undifferentiated 'space' into 'a place'". This is not Liz's place, this is the place constructed by adults, who by their very nature of being have very different ways of understanding the world than children or young people do. Liz is in transition between these two states of being, betwixt the worlds of childhood and adulthood. As identified by Valentine (2003, p.38) this boundary crossing when "childhood ends and adulthood begins is obscured by the luminal period of youth"; this is part of an ongoing process that shapes and defines a person. I believe that if we are going to attempt to understand what happens during this transitional phase, it is best to adopt a boundary-crossing approach to research, drawing from an interdisciplinary perspective that enables us to use the theoretical concepts and research methods that are most appropriate to the questions we want to address. Taking care as a specific example: this is a concept that invites a holistic approach; as a written word it may be small, and appear insignificant, however in its application and practice, it is multi-faceted and complex, resulting in a variety of potential interpretations. To explore this within the boundaries of one particular discipline would necessitate ignoring the very nature of it, limiting our ability to understand and make sense of it. However, it is important to note that this is an approach that invites critique and reproach, as it is widely acknowledged that "integrating the findings from different methods or different disciplines in anything other than the most superficial of ways is challenging" (Kessel et al 2009, p.32). My response to this is that this can challenge can be best addressed by presenting findings in a storied form, which maintains the complexity, avoids the "implicit reductionism in unidisciplinary work" (ibid., p.37) and allows different interpretations depending on the perspective (discipline and preference) of the reader.

Life is an innately emotional and spatial process (Jones 2005). How can we effectively provide support during youth, when as articulated by Jones (2008) we are forced to acknowledge that, as adults, even though we were once young people, we can no longer fully understand the world of young people. I have to acknowledge I am rapidly approaching my mid-century, and as such I am very different to Liz; I have all the baggage of adulthood with me. Recognising this difference serves to emphasise the importance of actively listening, of attending to, empathising with and caring about what Liz is saying, and also to the way I respond. I need to be conscious of normative, adultist expectations of how young people should be, how they should care, and what they should care about.

Relating this to my research: if we want to encourage young people to care about nature, and to demonstrate this through practical ways of caring for nature, we need to allow them to do it their way, not ours. And this requires us to take time to notice how they do this, for example Liz's slow, easily-missed moonlit 'wow': I noticed it because I was paying close attention, I was sharing the experience and ready to respond. This is all part of the caring process. I share this with you, the readers of this chapter with the aim of providing something more meaningful than a simple report of my research observations. With regard to Liz, this is a call to witness, to "share and deeply empathize with pain and suffering – the negative (although it could be applied to joy and love – the positive) and otherness – without fully knowing it" (Jones 2008, p.206).

Research presents us with a number of epistemological, methodological, ethical and political challenges (Jones 2008), particularly when it involves those that are 'other' than us (for example, bringing together those identified as children and young people or 'care-leavers' with those of us who are not). However, it also presents us with a number of ethical and moral challenges which need to be addressed. Some of these challenges can be viewed as problems, in that after due consideration it becomes clear what action should be taken (Banks 2010); others are more problematic and present as a dilemma in that whatever action is chosen, it involves some breaching of ethical principles. The choice is based on ethical reflection that includes a full contemplation of our morals and values, involving emotional as well as cognitive processes. As elucidated by Banks (2010, p.12): "ethical reflection is a process of mulling over, questioning and appraising a matter relating to human, animal or planetary flourishing". This combination of personal responsibilities (for example, codes of practice and ethical guidelines) together with personal values (and moral codes) is fundamental to the process of caring – we need to be moral to respect the wellbeing of those in a caring relation. It is not just about knowledge, it is more about receptivity, feeling and sensitivity (Noddings 2013) which contribute to awareness and understanding. From this stance we can work from an informed perspective that enables us to be more effective in supporting others to learn – and to learn ourselves – to be kind to ourselves and each other. Fables are one way by which we can communicate this in a way that bears witness to experience, respects otherness and conveys a moral lesson: they are a method for demonstrating care.

Narrated stories, such as the one within this chapter, are a vital way to develop our awareness of the ways in which different caring practices are experienced. The storied form allows us to engage with the emotional dimension of the experience, placing the central emphasis on the young person, on how Liz perceives himself and his position, his place in the world. The interdisciplinary nature of eliciting, interpreting and presenting research in this manner allows us to explore the experience in a more complex and holistic way, enabling us to learn so much more from it than would be possible from a more singular approach. By then embedding this discussion within a wider discourse that explores the concepts of care and caring, the focus shifts away from Liz and towards us, and to what we can learn from this, and meaningfully apply to our own situations – to our place(s) in life. So who should care about others, about nature? We *all* should, because as explained so wonderfully by Spretnak (2011, p.4), "all entities in the natural world, including us, are *thoroughly relational beings of great complexity, who are both composed of and nested within contextual networks of dynamic and reciprocal relationships*".

Within one of my final conversations with Liz I asked him what would be his best outing outside, where would he choose to go, and he replied: "*My Perfect Place? I would go where the moon is rising, just sit there and look at the surroundings.*" At the very heart of my PhD research is a commitment to care about how things are experienced, articulated and interpreted, so that what we offer to young people is effective at meeting their needs, as well as our own and the needs of what we call 'nature'. I recognise that this commitment causes me to get 'bothered' and 'troubled', and at time to get angry; but that is because I *care*. However, I now recognise that as well as getting angry, I have found a lot of humour in my research; there have been silly moments, funny moments, times shared with young people when we have laughed so much we have struggled for breath, magic moments when the mutual recognition of the absurdity of a situation has left us speechless, yet noisy with laughter. I feel the need to go back and revisit these 'fragile' moments (Macpherson 2008, p.1083) in a more mindful and careful way.

Writing this chapter has caused me to pause, reflect and extract meaning from this research encounter; this process has enabled me to reconsider my research from a number of different perspectives. Although there has been movement in recent years towards adopting a more interdisciplinary, creatively interpretive approach to research, this is still seen as controversial, arguably undisciplined, and is not generally accepted by policy makers as a credible method. There is still a political preference for more traditional, quantifiable and, in my opinion, simplistic methods, which ignore (or at the very least limit) the complexity, the nuance, the *messiness* and *funniness* of what we are studying. I find this unethical and more to the point, unkind to those we are studying. This is an area that warrants further research and publication: we need to keep up the momentum of challenge, and to promote more caring, humane ways to conduct and present research. As stated in the introduction, and argued throughout this discussion, kindness is a behaviour defined by ethical characteristics, demonstrated by acting in a friendly and considerate manner. Therefore, I end by kindly inviting you to pause once more, to remember Liz, and to consider what does care mean to you – and why?

Acknowledgements

Thank you to all the people, young and older, who have participated in my research, and to those who have guided me. A particular thank you for the kindness of the reviewers of the first draft of my chapter who have helped me to refine it, and provided me with inspiration for future foci of study.

References

Bachelard G. (1994) *The poetics of space*. Boston, Beacon.

Banks, S. (ed.) (2010) *Ethical issues in youth work*. London, Taylor and Francis.

Bray, L., O'Brien, M.R., Kirton, J., Zubairu, K. and Christiansen, A. (2013) The role of professional education in developing compassionate practitioners: a mixed methods study exploring the perceptions of health professionals and pre-registration students. *Nurse Education Today* 34, pp.480–6.

Carson, R. (1956) *The sense of wonder*. New York, Harper and Row.

Carson, R. (1962, reprint 2000) *Silent spring*. London, Penguin.

Chawla, L. (1990) Ecstatic places. *Children's Environments Quarterly* 7, pp.18–23.

Convery, I., Corsane, G. and Davis, P. (eds.) (2012) *Making sense of place: multidisciplinary perspectives*. Woodbridge, Boydell Press.

Denzin, N.K. (2014) *Interpretive autoethnography*. London, Sage.

Denzin, N.K. and Lincoln, Y.S. (2011) *The Sage handbook of qualitative research*. London, Sage.

Descartes, R. (1989) *The passions of the soul*. Indianapolis, Hackett.

Dunlap, J and Kellert, S.R. (eds.) (2012) *Companions in wonder: children and adults exploring nature together*. Boston, MIT Press.

Ellis, C., Adams, T.E. and Bochner, A.P. (2011) Autoethnography: an overview. *Qualitative Social Research* 12, unpaginated.

Ellis, C. (2004). *The ethnographic I: a methodological novel about autoethnography*. Walnut Creek, Altamira Press.

Fuller, R.C. (2006) *Wonder: from emotion to spirituality*. Chapel Hill, University of North Carolina Press.

Hayes, T. (2013) Seeing the world through their eyes. Learning from a 5 1/2 year old, a rabbit and a boat ride with aunty. *Horizons* 63, pp.36–9.

Hayes, T. (2014) The challenges of social inclusion in outdoor education: can tortoise and hare learn together? In Varley, P. and Taylor, S. (eds.) *Being there: slow, fast, traditional, wild, urban, natural ... Proceedings from the 2nd International Adventure Conference*. Fort William, University of the Highlands and Islands, pp.43–54.

Hayes T. (2015) A playful approach to outdoor learning: boggarts, bears and bunny rabbits! In Evans, B., Horton, J. and Skelton, T. (eds.) *Play and recreation, health and wellbeing, Geographies of children and young people, vol. 9*. Springer, New York, pp.155–172.

Haywood, M. (2012) Viewing the emergence of scenery from the English Lake District. In Convery, A., Corsane, G. and Davis, P. (eds). *Making sense of place: muldisicplinary perspectives*. Martlesham, Boydell and Brewer, pp.23–32.

Horton, J. and Kraftl, P. (2006) What else? Some more ways of thinking and doing 'Children's Geographies'. *Children's Geographies* 4, pp.69–98.

Hustedde, R. and King, B. (2002) Rituals: emotions, community faith in soul and the messiness of life. *Community Development Journal* 37, pp.338–48.

Jones, O. (2000) Melting geography: purity, disorder childhood and space. In Holloway, S. and Valentine, G. (eds.) *Children's geographies: living, playing learning*. London, Routledge, pp.29–47.

Jones, O. (2003) 'Endlessly revisited and forever gone': on memory, reverie and emotional imagination in doing children's geographies. *Children's Geographies* 1, pp.25–36.

Jones, O. (2005) An ecology of emotion, memory, self and landscape. In Davidson, J., Bondi, L. and Smith, M. (2005) (eds.) *Emotional geographies*. Aldershot, Ashgate, pp.205–18.

Jones, O. (2008) 'True geography [ ] quickly forgotten, giving away to an adult-imagined universe'. Approaching the otherness of childhood. *Children's Geographies* 6, pp.195–212.

Kessel, A., Green, J., Pinder, R., Wilkinson, P., Grundy, C. and Lachowycz, K. (2009) Multidisciplinary research in public health: a case study of research on access to green space. *Public Health* 123, pp.32–8.

Kraftl, P., Horton, J. and Tucker, F. (2014) *Children's geographies*. Available at: www.oxfordbibliographies.com/obo/page/childhood-studies

Lorde, A. (1984) *Sister outsider: essays and speeches by Audré Lorde*. Berkeley, Crossing Press.

Louv, R. (2005) *Last child in the woods*. Chapel Hill, Algonquin.

Macartney, B. (2007) What is normal and why does it matter? Disabling discourses in education and society. *Critical Literacy* 1, pp. 29–41.

Macpherson, H. (2008) 'I don't know why they call it the Lake District, they might as well call it the rock district!' The workings of humour and laughter in research with members of visually impaired walking groups. *Environment and Planning D* 26, pp.1080–95.

Marschall, A. (2014) Who cares for whom? Revisiting the concept of care in the everyday life of post-divorce families. *Childhood* 21, pp. 517–31.

Meyer, J. and Land, R. (2003) *Threshold concepts and troublesome knowledge: linkages to ways of thinking and practising within the disciplines*. Edinburgh, ETL Project.

Milligan, A. and Wood, B.E. (2009) Conceptual understandings as transition points: making sense of a complex social world. *Journal of Curriculum Studies* 42, pp.487–501.

Moules, N.J., Field, J.C., McCaffrey, G.P. and Laing, C.M. (2014) Conducting hermeneutic research: the address of the topic. *Journal of Applied Hermeneutics* 2014, article 7.

Nixon, J. (2014) The 'questionableness' of things: beyond method. Available at: www.yorksj.ac.uk/education--theology/faculty-of-etrs/faculty-news--events/faculty-events/previous-events/value-and-virtue-2014/conference-keynote speakers.aspx

Noddings, N. (2013) *Caring: a relational approach to ethics and moral education*. Berkeley, University of California Press.

OED [Oxford English Dictionary] (2011) *Oxford essential English dictionary*. Oxford, Oxford University Press.

Palmer, S. (2006) *Toxic childhood: how the modern world is damaging our children and what we can do about it*. London, Orion.

Pelias, R.J. (2004) *A methodology of the heart: evoking academic and daily Life*. Walnut Creek, Altamira Press.

Pyer, M. (2008) Unintended consequences? Exploring the un(fore)seen effects and outcomes of research. *Children's Geographies* 6, pp.213–17.

Ridgers, N., Knowles, Z. and Sayers, J. (2012) Encouraging play in the natural environment: a child-focused case study of Forest School. *Children's Geographies* 10, pp.49–65.

Rogers, R. (2011) 'I remember thinking, why isn't there someone to help me? Why isn't there someone who can help me make sense of what I'm going through?': 'Instant adulthood' and the transition of young people out of state care. *Journal of Sociology* 47, pp.411–26.

Ross, N., Renold, E., Holland, S. and Hillman, A. (2009) Moving stories: using mobile methods to explore the everyday lives of young people in public care. *Qualitative Research* 9, pp. 605–23.

Sigman, A. (2012) The impact of screen media on children: a Eurovision for parliament. In Clouder, C., Heys, B., Matthes, M. and Sullivan, P. (eds.) *Improving the quality of childhood in Europe*. Brighton, ECSWE, pp.88–121.

Sissay, L. (2012) *When I left care they said I was a great survivor*. Available at: www.theguardian.com/commentisfree/2012/oct/30/leaving-care-support-young-people

Sissay, L. (2013) *Why I'm holding a Manchester Christmas dinner for 45 people who were in care*. Available at: www.theguardian.com/uk-news/the-northerner/2013/nov/28/christmas-dinner-care-leavers-manchester-lemn-sissay

Smith, H. and Smith, M.K. (2008) *The art of helping others: being around, being there, being wise*. London, Jessica Kingsley.

Spretnak, C. (2011) *Relational reality: new discoveries of interrelatedness that are transforming the modern world*. Topsham, Green Horizon.

Sword, H. (2012) *Stylish academic writing*. Boston, Harvard University Press.

Valentine, G. (2001) *Social geographies: space and society*. Harlow, Pearson.
Valentine, G. (2003) Boundary crossings: transitions from childhood to adulthood. *Children's Geographies* 1, pp.37–52.
Waite, S. (2007) 'Memories are made of this': some reflections on outdoor learning and recall. *Education 3–13: International Journal of Primary, Elementary and Early Years Education* 35(4), pp.333–47.
White, R. and Stoecklin, V.L. (2008) *Nurturing children's biophilia: developmentally appropriate environmental education for young children*. Kansas, White Hutchinson Leisure and Learning Group.
Wood, B. (2007) Conflict, controversy, and complexity: Avoiding the 'slippery stuff' in social studies. *Critical Literacy* 1, pp.42–9.

12 Young women's care*ful* and care*less* drinking geographies

Samantha Wilkinson

Introduction

In this chapter, I draw on mixed-methods qualitative research conducted 'with' young people, living in Chorlton and Wythenshawe, Manchester, UK. I show that a care strategy predominantly talked about by young women under the legal drinking age in Wythenshawe and Chorlton involves sober, or less inebriated, friends supervising heavily intoxicated friends. Being a (relatively) sober minder is something young women commonly undertake; it can be pre-planned, or occur without being planned as a result of analysing the drunkenness and drunken behaviour of friends. For many young women, physical co-presence is important for their caring geographies. However, some young women make evident that caring relationships can transcend face-to-face acts. Digital technologies, such as mobile phones, can facilitate care about embodied experiences beyond one's immediate space and time. Technology does not mean that physical care is being supplanted by caring at a distance. Rather, technology supplements proximate care. Whilst friends expect that they should care for each other when drunk, the provision of care within friendship must be managed carefully, so as not to intrude on the autonomy of the friend who does the caring (MacLean 2015). However, equally, young women recognised that not acting in line with the role of 'carer' or 'protector' can be detrimental to friendship bonds.

Young people and care

Caring involves both 'caring for', for instance physical care, and 'caring about', including emotional concern about one's wellbeing (Silk 2000, Bowlby 2012). These are not mutually exclusive and, as Bowlby (2012) says, one often leads to the other. Further, care is not a one-way process; it usually involves reciprocity between the cared for and the carer. Research on the geographies of care has highlighted the importance of space and place to experiences of care (Milligan 2003, Dyck et al 2005, Bowlby 2012). As Bowlby (2012) asserts, much care requires co-presence, that is, to physically interact with the person being cared for. However, the growth in communication via the Internet and mobile phones offers new spaces of care that, as Bowlby (2012) argues, are not reliant upon co-presence.

In existing geographical literature on care, the corporeal bodies of the old have been privileged over those of young people (Atkinson et al 2011). This is not to say that there is any rigid distinction between the 'young' and the 'old'; indeed, authors have been keen to think of age in a more relational manner, for instance by considering intergenerational approaches to age (Hopkins and Pain 2007). Notwithstanding this, a predominant research focus on care for the old is problematic because, as Bowlby (2012) asserts, care is important to, and exchanged between, everyone – not solely those identified with state welfare provision. When the literature has considered young people as either being *cared for* or *caring*, it has typically been biased to the global South, for instance, Robson's (2004) work on young people caring for sick, elderly, dying or disabled adult family members in their Zimbabwe homes. A notable exception from the global North is Power's (2008) work on caring for young adults with intellectual disabilities; yet, the young people are portrayed as solely receiving care and their role as carers for others is not acknowledged. One of the few papers to recognise that the role of *cared for* and *carer* do not necessarily occur in isolation is Fine and Glendinning's (2005) work on young disabled people's experiences of being cared for, and also giving care in their roles as parents and wives. Whilst there has thus been a dearth of literature regarding how young people care for other people, work has considered how young people care for nonhuman entities, such as the environment (Bartos 2012).

A second critique of the existing geographical literature on care is that men's role as carers has largely been denied (Fine and Glendinning 2005). In focusing on caring as a 'feminine' activity, private places of the home have received significant attention as carescapes (Fine and Glendinning 2005), thereby largely neglecting public carescapes. With a focus on young women's drinking practices in a range of spaces, including outdoor drinkscapes, this chapter will critique these normative binary divides between 'male'/'female' and 'private'/'public' forms of care. As Roseneil and Budgeon (2004) assert, care increasingly takes place beyond the 'family', between networks of friends. According to the authors, friends give and receive care in a variety of situations of emotional, physical and practical need. Bowlby (2011, p.606) argues for a better understanding of the significance of friendship in geographies of care, arguing that an explicit consideration of friendship is excluded from the literature on caring geographies, "despite friendships' place as a source of care and support in many people's everyday lives". She suggests that the care involved in friendship is 'informal' care; it is not formally organised, and is typically unpaid. Further, caring between friends involves both giving and receiving, and there is an expectation of some kind of reciprocation. Bowlby (2011) warns that persistent inequalities in the exchange of care between friends may lead to the loss of the friendship.

Young people caring and drinking

Accounts of care in the geography literature construct care as typically enacted by a healthy person, towards people classified as 'ill' or 'unhealthy' (e.g. Andrews 2003, Liaschenko et al 2011, Wong 2014). However, this conceptualisation is restrictive and must be widened to take into consideration, for instance, how people 'care' for friends, or even, strangers who are experiencing the effects of alcohol. The paucity of research considering the embodied nature of young people's care for their drunken or hung-over friends or strangers is problematic because, as Lawson (2009) contends, everybody gives and needs care. There are, however, a few notable exceptions. Take Russell et al's (2011) qualitative study of young people's relationships with alcohol in County Durham and Darlington. The authors argue that friends look after one another by knowing the signs that limits are being reached, and will warn each other when they have reached their limits. In this vein, excessive drinking is seen as a risk to friendship, or means that young people are not able to care for their friends (Russell et al 2011). Further, Armstrong et al (2014) explore the range of protective behaviours young women deploy in public spaces in order to moderate the adverse effects of alcohol, arguing that most of these strategies derive from the friendship group to which the women belong. Armstrong et al (2014) argue that the friendship group offers protection against 'other drunken people', through what one participant identifies as 'safety in numbers'. 'Safety in numbers' is perceived as particularly important when in the presence of unknown males, or whilst travelling on public transport. Further, young women in Armstrong et al's (2014) study commonly mentioned monitoring and regulating the alcohol consumption of group members, whilst others recalled how they had given or received suggestions to slow down the rate of their alcohol consumption. According to Armstrong et al (2014, p.756), for women, more so than men, there is a "group-based culture of helping" in public space where alcohol is involved.

An additional example of this 'culture of helping' can be seen in Sinkinson's (2014) research into alcohol consumption in, on and around water in New Zealand. The young people in Sinkinson's (2014) study deploy strategies to keep themselves safe against the potential risks associated with combining aquatics and alcohol. These could be individual rules, such as: limiting the amount of alcohol they consume, or not drinking when participating in water activities; not swimming; or exercising more caution. Additionally, Sinkinson (2014, p.53) discusses "sober minders" as techniques deployed by young people to increase safety. Sober minders are the presence of people who could adopt the role of minders, along with having sober people outnumber intoxicated people. Despite drinking in high-risk spaces, young people in Sinkinson's (2014) study articulated strategies they utilise in order to maintain their own and others' well-being. However, there are barriers to the implementation of protective behaviours. Armstrong et al (2014) contend that heavily intoxicated group members may refuse to accept the supervision or instructions given by less intoxicated friends. More than this, the authors contend that maturity, individual differences in risk perception, and attitudes towards alcohol are factors influencing the extent to which one complies with group-based safety routines. It is also worth recognising that some friendship groups lack the "culture of helping" (Armstrong et al 2014, p.756) alluded to above. In such groups, as Armstrong et al (2014) recognise, there is a tendency to encourage, rather than manage, heavy or dangerous drinking amongst group members.

Moreover, Fjaer (2012), exploring interactions during hangovers amongst young Norwegian adults, illustrates that care between friends, when bound up with the consumption of alcohol, can extend beyond the drinking occasion. By spending the morning after the party together with friends, eating breakfast, discussing events, young people can provide support for those who cannot remember what happened, or are concerned about others' judgement of their behaviour. Fjaer (2012) illustrates that care practiced by friends towards each other is both physical – helping to obtain remedies, such as paracetamol and food – and emotional - helping to relieve binge-angst. 'Binge-angst' is a term used by Norwegian participants in Fjaer's (2012) study. In a typical scenario, one would wake up and think through the night before, and have trouble remembering what occurred. This leads to feelings of concern about having done something that might harm the impression others have of oneself. In line with Parr (2003, p.219, emphasis in original), I contend that there is something to be gained by interrogating the making of 'care*ful*' and 'care*less*' spaces.

Research context

I now provide a brief overview of the case study locations. Wythenshawe, a district eight miles South of Manchester city centre, was created in the 1920s as a Garden City in an attempt to resolve Manchester's overpopulation problem and 'depravation' in its inner-city slums. Wythenshawe continued to develop up to the 1970s. However, the 1980s and 1990s saw steady decline, high unemployment, decaying infrastructure, crime and drug abuse problems (Atherton et al 2005). The population of Wythenshawe increased from 66,000 in 2001 to over 70,000 in 2007 (Wythenshawe Town Centre 2013). Wythenshawe has commonly been labelled one of the largest council housing estates in England (*The Guardian* 2012). Wythenshawe wards have a resident white population that is above 90 per cent; this compares with 23.2 per cent of the population being from Black and minority ethnic communities across Manchester as a whole (Manchester City Council 2011). The area is thus dominated by white working-class drinking cultures. There are distinct neighborhoods within Wythenshawe, along with a town centre with various shops, supermarkets, hairdressers, pubs and a club. Wythenshawe is several miles from Manchester city centre, and faced with relatively poor transportation links (Lucas et al 2009).

Chorlton is a residential area approximately five miles from Manchester city centre, with a population of circa 13,500 people (Manchester City Council 2012). Chorlton is a cosmopolitan neighborhood with traditional family areas alongside younger, vibrant communities. The area has good road and bus access to, and from, the city centre, and is situated within easy access to the motorway network. Drawing on Manchester City Council's (2012) data: in 2011, pupils living in Chorlton achieved much higher educational results than the Manchester average. The percentage achieving five GCSEs A*-C was 69.5 per cent in Chorlton, compared to 57 per cent for Manchester's average. Chorlton has a higher proportion of minority ethnic residents in comparison to Wythenshawe, and compared to the national average (19.1 per cent cf. national average 11.3 per cent). As of November 2011, private residential property in Chorlton accounted for 90.3 per cent of all property in the ward, much higher than the city average of 68.7 per cent. In 2009, Chorlton had a much lower proportion of children under the age of 18 in poverty than the Manchester average (10.3 per cent compared to 39.9 per cent). However, Chorlton still has pockets of deprivation which are often underestimated by the council (Chamberlain 2012). Due to the increasing number of bars and restaurants in Chorlton, a pub watch scheme has been initiated (Young 2011).

40 young people were recruited for multi-stage qualitative research. In some respects, the sampling strategy was purposive, as I aimed to recruit 20 young people from each case study location, and aimed for an equal gender distribution. I recruited the majority of participants through gatekeepers at local schools, community organisations, youth clubs, and universities. In order to reach potential participants, I also distributed flyers and business cards to houses and businesses in both case study locations; posted on discussion forums concerning both areas; used Twitter and Facebook to promote my study to locals from each area; and arranged to be interviewed by the host of a local radio station. Some young people were initially cautious about participating in my study, due to worries about others finding out about their drinking practices. By building trust and friendship with participants (Valentine 2013), these participants could then tell their friends about the study and, from their first-hand experience, reassure friends that confidentiality and anonymity are strongly abided by; this is recognised as a snowballing sampling technique.

I had a palette of methods to utilise, and made it clear to the young people that they could 'opt in' to whichever method(s) they wished. As Holland et al (2008, p.19, emphasis in original) argue: "by enabling young people to choose *how* they wish to communicate with us we recognise them as social actors and begin to move our practice away from adult-centric procedures". The methods I draw on in this chapter include in-depth interviews, peer interviews and written drinking diaries. Many of these methods are deemed to be "child-friendly" (Punch 2002, p.321). Child-friendly methods are commonly employed in research with young people (see, for instance, Harris et al 2015), perhaps reflecting that the terms 'children' and 'young people' are often used interchangeably (Weller 2006). However, this conflation is problematic; it overlooks that young people have different experiences of space and place compared to children. The uniqueness of young people's lives means, just as with children, a special suite of methods should be in place to explore their lives. Recognising this, I complemented the above-mentioned methods with 'go along' mobile phone interviews – which involved asking young people to take photographs and videos during their nights out/in using their mobile phones, and, at a later date, to navigate through their mobile phones using photographs and videos as oral catalysts to tell me about their drinking experiences. This novel and fun method was positively received by young people, and proved to be a culturally credible means of researching their alcohol-related geographies.

With regard to analysing interviews, peer-interviews and diaries, I adopted the manual method of coding by pen and paper, perceiving that computer-assisted qualitative data analysis distances researchers from the data (Davis and Meyer 2009). Initially, following Miles and Huberman's (1994) three-stage model, a process of data reduction occurred, whereby I organised the mass of data and attempted to meaningfully reduce this. Second, I undertook a continual process of data display in the form of a table. Third, I undertook a process of conclusion drawing and verification. With regard to analysing sketch photographs/videos, emphasis was placed on the narratives of participants accompanying their products, in the form of 'go along' mobile phone interviews. This chimes with Barker and Smith's (2012) contention that the interpretation of images should be undertaken with participants to ensure that their intended meanings are explored, rather than interpretive meanings given by the researcher. These data forms were thus also analysed utilising Miles and Huberman's (1994) aforementioned procedure. Participants feature in this chapter through pseudonyms, so as to conceal their identities. Yet, in order to contextualise quotations, ages and locations are given.

'Sober' minders

A care strategy predominantly talked about by young women under the legal drinking age in Wythenshawe and Chorlton, involves sober, or less inebriated, friends supervising heavily intoxicated friends (see Armstrong et al 2014). This strategy can be illustrated by the following comments from Heather, Stacey and Kelly, below:

> So like, the most amount of people I'd probably go with is like eight people, that is it. I wouldn't go any higher than that because then I'd probably just worry about losing them. But I make sure it's in the two times table if you know what I mean. Like it's either two, four, six, or eight people, so then we can pair up and look after like one person. Cos if there's seven of us that means that there's gonna have to be a group of three, and two groups of two. So like make sure there's pairs of two so like we can look after each other. We'll tell one person, "Which one of you wants to get more drunk, and which one wants to stay sober to look after one of us?" Like he'll say "I'll get more drunk" and the other one will say "I'll look after them". So there'd be two, and one of them would be sober so we'd know what we'd be doing. So like, I think that's sensible. (Heather, 15, Wythenshawe, interview)
>
> Here's all of us together [pointing to a photograph of all her friends sat together at a house party], but sometimes only a few of us will drink, but we will more or less sit together, with a couple of sober people who will look after us. So it's really a close-knit group of friends I've got. So we all know each other's limits more or less and we'll all keep an eye on each other. (Stacey, 16, Chorlton, mobile phone interview)
>
> I stop drinking when I've got to look after somebody else or like if I'm at a party and there's loads of people there like I'll just put them in bed and then I'll go and check on them all the time. (Kelly, 17, Wythenshawe, peer interview)

As the above quotations suggest, being a (relatively) sober minder is something young people commonly undertake; it can be pre-planned, or occur without being planned as a result of analysing the drunkenness and drunken behaviour of friends. For instance, whilst Heather and Stacey explain that the 'buddy system' is agreed in advance, for Kelly it is less planned, and more tacitly encoded in mutually understood rules of friendship (see MacLean 2015). The buddy system/use of a sober minder was not something young people typically spoke negatively about, or objected to in any way. These findings contradict many conclusions drawn in the existing alcohol studies literature. For example, MacLean (2015) concludes that many young people dislike going out with friends who do not drink. To the contrary, my findings suggest that the use of a (relatively) sober minder was a common sense tactic. This was also the case with Sinkinson's (2014) research into young people's strategies for keeping safe when combining alcohol consumption and water activities in New Zealand. The author found that the presence of a sober person, or people who could perform the role of minders, and having sober people outnumber drunken people, were strategies young people commonly utilised to increase safety. Here then, my findings lend credence to the use of the term "calculated hedonism" (Szmigin et al 2008, p.359), a term recognising that young people manage their pleasure around alcohol.

Care at a distance

For the young women we have heard from so far, physical co-presence is important for their caring geographies (see Bowlby 2011). However, some young women in my study make evident that caring relationships can transcend face-to-face acts. Take the comment from Stacey below:

> Josie who was the first one I got drunk with she is now 18, so I gave her money and she bought me alcohol, after making sure that one of my friends was there to supervise me, and wouldn't be drinking. Calling up my friends every single hour to make sure I was okay, and ensuring that I would be fine … I get texts in the middle of the night saying "drink water". (Stacey, 16, Chorlton, interview)

In line with the findings from Armstrong et al's (2014) study, Stacey expresses a sense of safety derived from the knowledge that she is looked after by her friendship group. The notion of what Armstrong et al (2014, p.758) term a "culture of helping" is evident here. Interestingly, Stacey enrols the mobile phone into her consumption network; it is fundamental for facilitating an affective engineering of her experiences of drunkenness (see Duff 2012). Josie would call Stacey's friends to ensure she is "okay", alongside sending her text messages to ensure she is drinking plenty of water. Digital technologies, such as mobile phones, can thus facilitate care about embodied experiences beyond one's immediate space and time (Farrugia et al 2015). From this, it is evident that technology does not mean that physical care is being supplanted by caring at a distance. Rather, technology supplements proximate care; the caring gaze mediated through technology enables bodily transformations in "real space" (see Parr 2002, p.85).

Managing care when drinking

Whilst friends expect that they should care for each other when drunk, the provision of care within friendship must be managed carefully, so as not to intrude on the autonomy of the friend who does the caring (MacLean 2015). This was something commonly expressed by young women in my study, from both Chorlton and Wythenshawe. Take the following comments:

> It ruins your night though if someone, if one of your mates gets too drunk, cos then you have to look after them. (Vera, 15, Wythenshawe, peer interview)
>
> A couple of people have drank too much, and they're just kinda annoying, cos they're like, you have to look after them, and it's like "I don't want to, I don't want to look after you, it's your fault". (Amy, 18, Chorlton, interview)

As the above quotations from Vera and Amy indicate, some young people judge the behaviour of friends who drink too much (see de Visser et al 2013). Both Vera and Amy appear to resent it when one of their friends gets "too drunk" because caring for them can restrict their social and spatial freedom on a night out. However, equally, young people recognised that not acting in line with the role of 'carer' or 'protector' can be detrimental to friendship bonds. Consider the diary extract from Louisa below:

> Kate and I had 2 cocktails at The Dainty [a bar in Manchester city centre], just chatting and catching up. Then we went to a cheap bar on Wayward street for cheap drinks (2 shots and 2 double mixers each). We stayed there for about half an hour then decided to go back to a house party in Chorlton. We bought a bottle of wine for the bus journey back. Kate drank most of the wine and was wayyyy to drunk by the time we got to the party: my friends helped me look after her in the bathroom until she passed out on a bed for about half an hour. Everyone at the party was taking drugs (not drinking much) and I took some MDMA[1] whilst Kate was asleep. When she woke up I had to help carrying her home, made sure she was asleep and went back to the party (BAD FRIEND!). (Louisa, 22, Chorlton, diary)

Here, Louisa describes acting under the influence of alcohol and drugs whilst attempting to look after her intoxicated friend, Kate. Whilst performing some caring strategies, such as getting her drunken friend home, Louisa describes herself as a "BAD FRIEND!" for leaving Kate whilst she resumed partying. Louisa seems to recognise, in hindsight, that care is very important to friendships. This supports contentions made in the literature. For instance, in relation to drinking, MacLean (2015) argues that being let down by a friend is distressing, and can damage friendship bonds. Further, Niland et al (2013) contend that friends who do not stay together when sharing drinking fun are negatively positioned.

Concluding remarks

Public debate in the UK is permeated by a rhetoric of anxiety relating to 'out-of-control' young people (Parkes and Conolly 2011), who are positioned as lacking capacities of/for self-regulation (Kelly 2003). The examples of young people's gestures of care and responsibility for each other on nights out, shown throughout this chapter, contrast markedly with popular representations of 'anti-social' young people in public spaces (see Brown 2013). As I have shown throughout this chapter, friendship was a fundamental constituent of care strategies talked about by the young women in my study. There was a "culture of helping" (Armstrong et al 2014, p.758) in many female friendship groups. For instance, many young women under the legal drinking age have a 'buddy system', in which a sober, or less inebriated, friend supervises an intoxicated friend. Young women could pre-plan this system, or enrol it during the course of a night out, by analysing the drunkenness and drunken behaviour of friends. Friendship caring strategies did not always rely on physical co-presence. Young people often enrolled mobile phones in consumption networks, enabling care about embodied experiences beyond one's immediate space and time (Farrugia et al 2015). Young people also showed how alcohol can present risks to friendships; not acting in line with the role of 'carer' could result in being labelled a 'bad friend'. Equally though, tensions within a friendship group may arise when a drunken friend intrudes on the autonomy of the friend who does the caring (MacLean 2015). Through these findings, I have contributed towards a better understanding of the significance of friendship in geographies of care (Bowlby 2011). I urge scholars to conduct research which explores the emotional and embodied nature of care and young people's caring practices, whilst critiquing normative representations of care.

Acknowledgment

The research reported in this chapter was funded by both an Economic and Social Research Council North-West doctoral Training Centre CASE +3 award [ES/J500094/1] and an Alcohol Research UK Postgraduate Research Studentship [RS 12/02].

Note

1 MDMA is commonly known as ecstasy and, more recently, 'Molly'. It is a synthetic, psychoactive drug that has similarities to both the stimulant amphetamine and the hallucinogen mescaline. It produces feelings of increased energy, euphoria, emotional warmth and empathy toward others, and distorts sensory and time perceptions (NIH 2013).

References

Andrews, G.J. (2003) Locating a geography of nursing: space, place and the progress of geographical thought. *Nursing Philosophy* 4, pp.231–48.

Armstrong, K., Watling, H., Davey, J. and Darvell, M. (2014) A qualitative exploration of young women's drinking experiences and associated protective behaviours. *Journal of Youth Studies* 17, pp.749–62.

Atherton, J., Baker, C. and Graham, E. (2005) A 'genius of place'? Manchester as a case study in public theology. In Graham, E. and Rowlands, A. (eds.) *Pathways to the public square: proceedings of the 2003 IAPT Conference, Manchester UK*. London, Transaction Publishers, pp.63–83.

Atkinson, S., Lawson, V. and Wiles, J. (2011) Care of the body: spaces of practice. *Social and Cultural Geography* 12, pp.563–72.

Barker, J. and Smith, F. (2012) What's in focus? A critical discussion of photography, children and young people. *International Journal of Social Research Methodology* 15, pp.91–103.

Bartos, A.E. (2012) Children caring for their worlds: the politics of care and childhood. *Political Geography* 31, pp.157–66.

Bowlby, S. (2011) Friendship, co-presence and care: neglected spaces. *Social and Cultural Geography* 12, pp.605–22.

Bowlby, S. (2012) Recognising the time-space dimensions of care: caringscapes and carescapes. *Environment and Planning A* 44, pp.2101–2218.

Brown, D.M. (2013) Young people, anti-social behaviour and public space: the role of Community Wardens in policing the 'ASBO generation'. *Urban Studies* 50, pp.538–55.

Chamberlain, V. (2012) *Council putting Chorlton to the back of the queue*. Available at: http://victorchamberlain.blogspot.co.uk/2012/04/council-putting-chorlton-to-back-of.html

Davis, N.W. and Meyer, B.B. (2009) Qualitative data analysis: a procedural comparison. *Journal of Applied Sport Psychology* 21, pp.116–24.

De Visser, R.O., Wheeler, Z., Abraham, C. and Smith, J.A. (2013) 'Drinking is our modern way of bonding': young people's beliefs about interventions to encourage moderate drinking. *Psychology and Health* 28, pp.1460–80.

Duff, C. (2012) Accounting for context: exploring the role of objects and spaces in the consumption of alcohol and other drugs. *Social and Cultural Geography* 13, pp.145–59.

Dyck, I., Kontos, P., Angus, J. and McKeever, P. (2005) The home as a site for long-term care: meanings and management of bodies and spaces. *Health and Place* 11, pp.173–85.

Farrugia, D., Smyth, J. and Harrison, T. (2015) Affective topologies of rural youth embodiment. *Journal of the European Society for Rural Sociology* 56, pp.116–32.

Fine, M. and Glendinning, C. (2005) Dependence, independence or inter-dependence? Revisiting the concepts of 'care' and 'dependency'. *Ageing and Society* 25, pp.601–21.

Fjaer, E.G. (2012) The day after drinking: interaction during hangovers among Norwegian adults. *Journal of Youth Studies* 15, pp.995–1010.

Harris, C., Jackson, L., Mayblin, L., Piekut, A. and Valentine, G. (2015) 'Big Brother welcomes you': innovative methods for research with children and young people outside of the home and school environment. *Qualitative Research* 15, pp.583–99.

Holland, S., Renold, E., Ross, N. and Hillman, A. (2008) *The everyday lives of children in care: using a sociological perspective to inform social work practice*. Available at: http://eprints.ncrm.ac.uk/466/1/0108%2520everyday%2520lives%2520of%2520children.pdf

Hopkins, P. and Pain, R. (2007) Geographies of age: thinking relationally. *Area*, 39(3), pp.287–294.

Kelly, P. (2003) Growing up as risky business? Risks, surveillance and the institutionalized mistrust of youth. *Journal of Youth Studies* 6, pp.165–80.

Lawson, V. (2009) Instead of radical geography, how about caring geography? *Antipode* 41, pp.210–13.

Liaschenko, J., Peden-McAlpine, C. and Andrews, G.J. (2011) Institutional geographies in dying: nurses' actions and observations on dying spaces inside and outside Intensive Care Units. *Health and Place* 17, pp.814–21.

Lucas, K., Tyler, S. and Christodoulou, G. (2009) Assessing the 'value' of New Transport Initiatives in deprived neighbourhoods in the UK. *Transport Police* 16, pp.115–22.

MacLean, S. (2015) Alcohol and the constitution of friendship for young adults. *Sociology*, online early, http://journals.sagepub.com/doi/abs/10.1177/0038038514557913 (accessed 21 Feb 17).

Manchester City Council (2011) *Childcare Sufficiency Assessment District Report: Wythenshawe*. Manchester, Manchester City Council.

Manchester City Council (2012) *Chorlton Ward Profile*. Manchester, Manchester City Council.

Miles, M.B. and Huberman, A.M. (1994) *Qualitative data analysis*. London, Sage.

Milligan, C. (2003) Location or dis-location? Towards a conceptualisation of people and place in the care-giving experience. *Social and Cultural Geography* 4, pp.455–70.

NIH (National Institutes of Health) (2013) *Drugfacts: MDMA (Ecstasy or Molly)*. Available at: www.drugabuse.gov/publications/drugfacts/mdma-ecstasy-or-molly

Niland, P., Lyons, A.C., Goodwin, I. and Hutton, F. (2013) 'Everyone can loosen up and get a bit of a buzz on': young adults, alcohol and friendship practices. *International Journal of Drug Policy* 24, pp.530–7.

Parkes, J. and Conolly, A. (2011) Risky positions? Shifting representations of urban youth in the talk of professionals and young people. *Children's Geographies* 9, pp.411–23.

Parr, H. (2002) New body-geographies: the embodied spaces of health and medical information on the internet. *Environment and Planning D* 20, pp.73–95.

Parr, H. (2003) Medical geography: care and caring. *Progress in Human Geography* 27, pp.212–21.

Power, A. (2008) Caring for independent lives: geographies of caring for young adults with intellectual disabilities. *Social Science and Medicine* 67, pp.834–43.

Punch, S. (2002) Research with children: the same or different from research with adults? *Childhood* 9, pp.321–41.

Robson, E. (2004) Hidden child workers: young carers in Zimbabwe. *Antipode* 36, pp.227–48.

Roseneil, S. and Budgeon, S. (2004) Cultures of intimacy and care beyond 'the family': personal life and social change in the early twenty-first century. *Current Sociology* 52, pp.135–59.

Russell, A., Lewis, S., Matthijsse, M. and Masson, K. (2011) *Revealing alcohol narratives: a qualitative study of young people's relationships with alcohol in County Durham and Darlington*. London, Public Health England.

Silk, J. (2000) Caring at a distance: (im)partiality, moral motivation and the ethics of representation. *Ethics, Place and Environment* 3, pp.303–9.

Sinkinson, M. (2014) Having a good time: young people talk about risk and fun when combining alcohol consumption and water activities. *International Journal of Health Promotion and Education* 52, pp.47–55.

Szmigin, I., Griffin, C., Mistral. W., Bengry-Howell, A., Weale, L. and Hackley, C. (2008) Re-framing 'binge drinking' as calculated hedonism – empirical evidence from the UK. *International Journal of Drug Policy* 19, pp.359–66.

The Guardian (2012) *On the white bus to Wythenshawe – council housing by design*. Available at: www.theguardian.com/commentisfree/2012/may/22/white-bus-to-wythenshawe-council-housing

Valentine, G. (2013) Using interviews as a research methodology. In Flowerdew, R. and Martin, D. (eds.) *Methods in human geography*. London, Routledge, pp.110–27.

Weller, S. (2006) Situating (young) teenagers in geographies of children and youth. *Children's Geographies* 4, pp.97–108.

Wythenshawe Town Centre (2013) *Real community – real lives Wythenshawe – Wythenshawe town centre making a forgotten dream a reality*. Available at: www.wythenshawetowncentre.co.uk/news-and-events/index.php/2013/10/real-community-real-lives-wythenshawe-wythenshawe-town-centre-making-forgotten-dream-reality/

Wong, M. (2014) Geographies and strategies of caregiving among skilled Ghanian migrant women. *Women's Studies International Forum* 42, pp.28–43.

Young, H. (2011) *Manchester: responsible alcohol sales project*. Available at: www.alcohollearningcentre.org.uk/LocalInitiatives/projects/projectDetail/?cid=6495

13 Views of young people with cognitive disability[1] about care in their relationships

Sally Robinson, Karen R. Fisher, Malcolm Hill and Anne Graham

Introduction

The relationships of young people with cognitive disability are often only considered through a lens that positions the young person as the care recipient. Research and policy has traditionally approached them as recipients of care, predominantly framed in a passive light. Theories of care that recognise mutuality in relationships acknowledge that such positioning evokes a shallow view of care. More recent research which takes a participatory or inclusive focus, working in collaboration with young people, presents a more complex and contested picture. To date, very little research has sought the perspectives of young people with cognitive disability about how they understand and experience care in their relationships.

The gap between traditional care research and young people's experience is highlighted by the conceptual approach of the most recent United Nations Convention on the Rights of Persons with Disabilities 2006. This frames disability as a social experience within an interrelational context (Kayess and French 2008). It explicitly recognises the agency of people with disability in personal, family and community relationships, including care relationships. This chapter explores the question of how young people experience care and caring by drawing from research that used participatory methods to articulate the patterns of connection to place and people for young people with cognitive disability living in regional Australia. Specifically, the research sought to better understand what helped young people with cognitive disability to feel a sense of belonging and connection in their regional communities, and what hindered this. Relationships of care and caring were dominant themes that emerged in the research, highlighting more nuanced understandings and experiences, which we explore in this chapter.

Care, caring and young people with cognitive disability

This article addresses caring relationships in several senses. A fundamental distinction is commonly made between caring about someone (loving or being fond of them) and caring for them (looking after them in some way) (Nolan et al 1996). Crudely this is a distinction between an emotional or affective relationship and a practical one, though the two can be of course combined. Similarly, carers may focus on only one of these or both. Further subdivisions may be made, e.g. practical caring for someone may include any of feeding, transporting, organising and arranging, advising, teaching etc. Another possible meaning of care arises when we speak of caring about the reactions of others, whether or not they are people we know well. Thus some people will be fairly indifferent to what others think, but as in our sample many are very sensitive and take to heart hurtful actions and statements of other people. People tend to care about exclusion from places or groups that are attractive to us or potentially useful for us (Hall 2005, 2010, Hill 2002).

Existing literature on care and caring of young people with disability broadly clusters around five areas: young people as receiving care (Power 2008, Rapanaro et al 2008, Schneider et al 2006); young carers (Day 2015, Hamilton and Adamson 2013, Smyth et al 2011); young people's experience of relationships (Jahoda and Pownall 2014; Anderson et al 2011); theories of care (Honneth 1995; Fitzgerald et al 2010); and social geographical framings of care (Hall 2005, 2010, Power 2013). While there is not scope to examine all these in detail here, the following provides a broad overview of each to draw attention to the ways in which young people with disability are often differently positioned in relation to care in the literature.

To date, much disability research has focused on the care duality between the person with disability as cared for and the family members providing the care (e.g. Wodehouse and McGill 2009). More recently, the research has diversified to also focus on young people's participation in care, with increased conceptualisation of agency and participation of young people in their relationships with families and friends, including relationships of care (Anderson et al 2011, Day 2015). A growing body of literature has also focused on young people as carers (Smyth et al 2011), but does not dwell on the experiences of young people with disability as active carers with other family members. It does note that young people who are carers are more likely to have a disability themselves. The evidence about young carers highlights the extraordinary levels of care of a family member with illness or disability not expected of other children and young people (Hamilton and Adamson 2013). It does not focus on the more general conceptualisation of any young people's contribution to family life as a mutual experience of care or caring reciprocated between family members.

Recent research on how young people with cognitive disability view relationships emphasises that, like other people, they seek security, equality and respect within friendships (Sullivan et al 2016). The findings demonstrate the importance of relationships not only for the social and practical elements, but also for developing a deeper sense of self, particularly around identity and belonging. Positive relationships can become a space where people experience *recognition*, contributing to a sense of self (Honneth 1995). Sullivan et al's research with young people about friendships found that they experienced barriers that inhibited positive relationships, including social and physical isolation that restricted the places to encounter others and limited valuing and support within relationships, limiting their agency.

The research reported in this chapter builds on Sullivan et al's (2016) findings by exploring care and caring within relationships of young people with cognitive disability, to gain a better understanding of whether, when and how young people

Table 13.1 Sullivan et al's conceptualisation of care in relationships

Sullivan et al conceptualisation	*As articulated in this chapter*
Relationships feel safe and useful	Utility in caring relationships
Control within relationships	Agency in caring relationships
Relationships enabling an ordinary life	Opportunity in caring relationships

navigate such barriers. Given that Sullivan et al's research is consistent with theories of care that emphasise mutuality and reciprocity (Honneth 1995, Fitzgerald et al 2010), our research also seeks to understand how young people with cognitive disability experience or view that mutuality of caring in their relationships. It does so by applying Sullivan et al's framework of whether relationships feel safe and useful; control within relationships; and relationships enabling an ordinary life, which we have conceptualised here as utility, agency and opportunity within caring relationships (see Table 13.1).

The research reported here draws from social geography, with its emphasis on mutually inhabited spaces, places and relationships, to help identify the ways in which care relationships are acted and enacted. This approach opens a space to think about the ways in which young people engage in reciprocal relationships of care (Power 2008, Hall 2010). It also potentially problematises the boundaries of care relations, from the perspective of young people, in this case those with cognitive disability, a group less likely to have their agency recognised, even within their intimate family relationships.

Space, Place, Relationships project

Our full project, *Space, Place, Relationships*, used participatory methodologies to explore with young people with cognitive disability how they understood belonging and connectedness in the regional (small town and rural) communities in which they lived. Drawing on concepts from social geography, the project aimed to better understand the patterns and importance of places and people to these young people, and the ways in which spaces (physical, social and cultural) supported and constrained their sense of belonging and connectedness to place and people. A three-stage design was developed which prioritised young people's participation through interviews, pictorial mapping, a photographic project based on photovoice and a survey.

Thirty young people in three country towns across three Australian states were involved in the core elements of the project. In each of the three sites, there were young people with a very diverse range of strengths, capabilities, support needs and experiences. Research relationships were developed with young people using existing community engagement and support programs. The qualitative research was conducted in three comparable regional centres in NSW, Victoria and Queensland. All three sites are defined as outer regional centres (more than

30km from a major city), and they have populations between 40,000 and 50,000 (ABS 2012).

Data from the first two methods were drawn upon to address the interests explored in this chapter, that being how young people with cognitive disability view care in their relationships. All names in the chapter are pseudonyms chosen by the young people, or for those with support needs such that they were not able to choose, by the researchers. The research methods emphasised accessibility in all stages – for example, interviews with all young people were supported by pictorial mapping, which helped foster conversation, keeping the young people as the expert about their lives, and leaving them with a detailed map of places and people they felt were important to them. This data provided a firm foundation upon which to start their photographic projects. The individual responses of young people to questions about what belonging and connection meant to them were confirmed (with permission) in group workshops, forming a grounded conceptualisation of belonging. Young people captioned and uploaded images to a private website, with support, during the project. All of the young people were involved in analysing their photographic data, by selecting the most important of their photographs in an interview and providing additional detail about the meaning and significance behind the images. Finally, the prioritised photographs were exhibited in public galleries as a concluding research activity – a way for us to publicly acknowledge, affirm and thank young people for their involvement, with these events proving to be highly validating for many of our co-researchers. A detailed description of the methods and findings of the full project can be found at Robinson et al (2014). As young people reflected on their connections, strong themes emerged about care and caring. These included capability, reciprocity and mutuality, recognition and tensions around agency and authority. Each of these themes is analysed below using Sullivan et al's framework on utility, agency and opportunity in the relationships of young people with cognitive disability.

Utility in caring relationships

Similar to Sullivan et al's research that found young people sought feelings of safety and usefulness within relationships, our study also pointed to the value young people with cognitive disability placed on stable, reliable and trustworthy relationships. These relationships were useful in their lives to achieve functional goals, shared with other people. Such findings reveal the instrumental dimension of care in the types of relationships they valued, and how these relationships reinforced recognition of their capabilities and mutuality.

Young people's relationships

Relationships with other people were of central importance to most of the young people. The great majority spoke explicitly about the importance of friends and family in their lives, articulating verbally and showing through their photography both the importance of having key relationships active in their lives, as well as the activities that they undertook with particular close people. Their connections through relationships with family, friends, and support workers (personal assistants) were articulated in key places – home, schools and disability support services, virtual worlds – as well as local public and semi-public places and spaces (such as shops and sporting groups). Many of these connections were similar to other young people in their community, but disability and rurality sometimes limited the breadth and depth of relationships for some of them. The impact of these interpersonal relationships was both positive and negative for many people.

Relationships with family dominated the young people's maps, photography and interviews. They spoke about their family, especially parents, as a key facilitator for activities, learning and practical and emotional support. In their photos, young people primarily described family life as oriented around shared daily routines of living and mutual support and care for one another. Younger people aged under 18 years, who were still living in their family homes, talked about patterns of family life that were similar to those of other young people their age in their locality. Older young people with lower support needs talked about increasing independence from family life in their social lives and activities, but most still lived at home.

Extended family, particularly grandparents, were also very important in young people's lives, with two thirds of participants including them on maps or in photos. Extended family seemed particularly important among the many young people who talked about loss or tension in their nuclear families, discussed below. The impact of these significant events and relations within their nuclear or extended families was profound, influencing other domains of their lives and identities. The young people mapped friendships and social connections in their community. Most of them named few people outside their families and service networks and many of these relationships appeared shallow and limited. The number and location of friends is not an indicator of how well or easily young people make friends, or on the quality of their friendships, as many people have small and intimate social networks. However, some young people named friends, at least in part, based on proximity due to shared service or school use, rather than shared interests, people they liked or experiences shared over time. For example, workers supporting the formation of relationships of people with higher support needs seemed to focus on planning and implementing service-related activities, individual support and the development of skills for increasing independence, more than planning for friendship development, in the research observations.

Many of the young people spoke about their feelings of loneliness and their desire for more friends. These feelings were also less directly present in the commentary of some others, who spoke about activities with family instead of with preferred friends, watching other young people and wishing to join them but not feeling confident, and feeling geographically isolated. For example, one young woman spoke about her feeling that people avoided speaking to her, misreading the signals that she sent out and thinking that she did not want to talk, when in fact she would like to. She arranged a photo of herself holding a sign saying, "you can talk to me". She said, "I just thought that a lot of people should know I'm not the big bad wolf!" Another young woman who had recently left school said she found it hard to get used to new people, and that she felt lonely during the week at the disability service at times: "I'm used to my normal friends." One participant wrote, "People don't understand me because I look ok, but I can't do most things so they just ignore me. It hurts." Another person directly linked loneliness to belonging, saying,

> It's not fun to feel you don't belong. Cause it's like everyone's over here [draws three people's heads close together] and you're all the way over here [draws a single head 10 cm away] with your arms crossed, and it's just like "why don't I belong?" cause you feel really sad [draws tears and sad face on the single head]. And everybody's all smiling and laughing without you [draws smiling faces on the clustered heads] and you're all the way over there crying.

The service context of some friendships may also have masked some loneliness, as referring to colleagues in services as friends precluded some conversations about friendship from occurring. Some of the young people distinguished friendships from work relationships and relations with other people using the same support services, but young people with high support needs were less likely to make the distinction. Relationships with disability support workers were important in the lives of many young people as a connection in itself and to facilitate other relationships. Service providers such as teachers and health professionals also featured in young people's narratives. Wider social relationships with neighbours, family friends, local business owners and shopkeepers, and taxi drivers formed a backdrop of looser social connection for many young people. These relationships appeared in photos and commentary and were observed on the visits to the localities.

Recognition and capability

From the perspective of the young people, one of the features of these relationships that facilitated belonging and connection was closeness, warmth and mutual positive regard within a relationship. Almost all participants in the study demonstrated that having other people viewing them as capable was very important, as evidenced through both their spoken and photographic contributions. Many of them were proud that they took responsibility and made active choices. Almost half the young people spoke about ways in which they took care of their own wellbeing, including strategies that they used to manage complex health conditions, exercising for wellbeing, and strategies to maintain fragile mental health.

Young people who were not able to articulate the importance of being seen to be good at something showed us repeatedly in their photos that recognising their capability mattered to them. Many statements on their collages centred on self-identified strengths and qualities, such as being athletic, artistic, a good cook, dancer or worker or taking care of others. This was particularly true of people with high support needs, who had fewer photos of relationships and more photos about capability. They had many statements about 'I can' or 'I do' accompanying their photos about what they achieve.

The young people who used a lot of disability support spoke directly about positive recognition in their relationships with support workers. They felt encouraged and motivated by support workers they liked, and this influenced their preferences and choices about what they wanted to do. In these positive support relationships, the depth of mutual knowledge of the young person and the worker of the character, preferences and communication styles of the other was evident in the research activities.

Leanne's example is illustrative of the importance that young people placed on demonstrating their skills. Leanne chose several photos of herself writing, a greatly preferred activity which she spends a lot of time doing, when she participated in the analysis interview. While she does not write words, she spends a lot of time practising and she is very motivated to write and type. She chose a photo of herself writing for her large image to be displayed in the art gallery, saying, "Leanne is writing. Writing a big job." At the end of the analysis interview, she looked at her selection, and said with evident satisfaction, "It's very, very good. Ok." Support workers who knew her well commented on the significance of writing in her life and the gravity with which she approached the task, saying that she did not like to be interrupted when writing and took considerable care with her work.

Some younger people had similar positive experiences in relationships with some teachers when they felt known and liked by their teachers and they spoke about feeling supported. The school environment sometimes compromised these opportunities for positive relationships between teachers and young people – for example, one student liked her maths teacher in particular, and said she wanted to learn, but found her class so disruptive that it was too hard. Open conflict with teachers and other students made school difficult and a negative place for two other young people. At the core of their negative feelings about school was that they felt misunderstood and disregarded by teachers and other students.

Agency in caring relationships

Utilising Sullivan et al's (2016) framework, we turn now to a second important theme emerging in our study around relationships and agency. Similar to Sullivan's research, we found that young people reflected that the quality and functioning of their relationships was affected by who was in charge, whether they had control or dependence, whether one or the other member of the relationship dominated or whether agency was shared. Their stories illustrated that their relationships could be enabling or disabling if their agency was supported or denied. Sometimes this denial of agency was benevolent in the name of protection; sometimes it was discriminatory, assuming no capacity. In contrast to the findings above about recognising strengths, the young people spoke of the injustice they felt when they were disempowered by the way they were treated in relationships in the name of disability.

Understanding this aspect of the framework is particularly relevant to the wider focus of this book, if relationships and acts of care and caring deny rather than enable agency. This study found that young people were concerned about reciprocity and mutuality in the relationships and liked support workers who recognised and facilitated agency in their relationships. Young people in this study did not position themselves as passive recipients of care, but as active and contributing family members, friends and partners in support relationships. Being an active agent in how relationships are enacted, in how support is provided and in choice and decision-making on individual issues was important to many young people. This was expressed through photos and commentary about making choices and decisions, being a friend, and having a say about the way the support role was carried out.

Reciprocity and mutuality

Many young people raised the importance to them of having reciprocity in their relationships with family members, friends and support workers recognised. Their perspective on reciprocity emerged from the data in multiple ways – through photographs and captions that showed young people providing practical support, such as meals and making statements of care for their families; through statements about how they care about their friends and show their concern; and through the depiction of family and friends as people of key importance in the maps of all young people, and support workers in most. When relationships were less reciprocal, some young people felt disregarded. One person noted, "Nobody listens to us", which was reflective of the views of a small number of participants.

Most of the young people in our study presented their relationships with family as foundational to their connections. Where young people described their family relationships as facilitating their sense of belonging and connection, they described or documented supportive and reciprocal relationships in which they actively contributed to the family. They explained that care and support in their family was mutual; for example, Matt said, "I love my family, and I try to look after them as much as I can." Some people had difficulties and tensions in friendships, which increased the importance of family relationships to them.

Several young people talked about home life being difficult due to conflict, loss or loneliness. A third of young people in the research talked about difficulties of various kinds that negatively affected the way they felt about family relationships. Some described a core of love and support, but within which they had a lot of arguments with parents and siblings. Others described repeated emotional and physical conflict within the family, feeling let down, not feeling able to trust family members, being taken into care, and being 'given a hard time' about core parts of their personal identity. Where young people felt their base of family support was not reliable, their confidence and optimism was affected.

Experience of reciprocity in friendships varied among the young people in the research. At its most meaningful, young people identified reciprocal elements in friendships that resourced and supported them. These included having shared interests, appreciating each other, giving to each other, being able and wanting to spend time together, supporting and being supported emotionally, and being able to survive a few bumps in the path. These relationships were expressed through comments such as, "good friends have your back at all times", "we do lots of things together" and "we've been friends a long time". However, many young people who did not have high service support needs talked about low-level interpersonal difficulties in their social networks, mostly driven by not having enough to do, gossip and negative commentary on Facebook. These negative experiences rumbled along as an undercurrent in the relations between young women in particular, which they referred to as 'drama'.

Facilitating agency in care relationships with support workers

Relationships with disability support workers were core in many of the young people's lives. Younger people still at school were less likely to have relationships in disability services and were more influenced by family. Most young people did not talk a great deal about the qualities of the relationships with their support workers, beyond limited descriptive terms. It was difficult for them to speak about unequal power relations, particularly for those who had a support person with them during the research activities, so these people were not asked directly about their relationships with support workers. It is difficult for research which is conducted at a distance to make too strong a determination about this balance in relationships between support workers and young people in this research.

Where young people had the most agency in their relationship with paid support workers, they had a clear conception of the role of their worker, the distinction between worker and friend, and also a close personal connection with their worker – both partners in the interaction had a warm regard for one another. For example, John Sena and Andrew worked together and each reflected on how they facilitated agency in the relationship.

JOHN SENA SAID OF ANDREW: "He is very friendly, and helps me on my music career."

ANDREW SAID OF JOHN SENA: "For me, this year with John Sena has been about learning to pull back and learn from watching him take the lead on our joint artistic projects."

A delicate balance between support that enables or constrains agency in personal development and relationships was evident. The warmth and rapport in ongoing relationships between some support workers and young people was clear to see, and many times, having support workers involved in the interviews and photographic processes gave reassurance to young people and gave helpful additional context to the researcher. However, the perspectives of the young people had less space to emerge, and approaches to their participation were more formulaic where the support workers were less focused on the agency and interests of young people or where a deficit approach prevailed. These findings were most evident in the observational logs and indirectly in the contributions of young people through their maps, photos and interviews.

Whether a particular worker facilitated agency within their relationship with the young person often seemed to directly affect the quality of the person's experience. For example, Zac included a photo of the whiteboard with names of workers and young people, which changes daily and allocates workers to the people they support for the day. When asked if he checked it when he arrived in the morning, Zac said,

> Yeah, it's a bit of a run-around thing. As soon as you walk in, it's like, "Who am I with?", going to the board … Yeah, yeah, it changes how I'm feeling about the day [depending on who I'm with].

Opportunity in caring relationships

The third and final theme we raise in this chapter concerns opportunity in caring relationships. We have conceptualised this theme from Sullivan et al's framework, which discusses how and whether relationships facilitated an ordinary life as defined by the young people, that is they felt valued and respected, gained independence, felt accepted and could form relationships like other young people in their community. Positive and supportive relationships were important for young people's feelings about being known, valued and equipped to engage with their community, such as Rachel's experience. Rachel included several photos from home about music and the ribbon she used for dancing, so it was clear she liked to dance. She also included a photo of dancing class with her support worker, Jane, teaching. The following interchange shows how much she is motivated by Jane's presence in the class:

> *Why do you like dancing, Rachel?* I like Jane. *Do you like what she's teaching you?* Yeah. She's nice. I like her. Jane's nice. I like her. Very nice. She's nice, that Jane. Yeah. *Is that the best thing about dancing?* Yeah. I like Jane. Jane's just there [pointing to photo].

When young people felt respected and support workers fostered their contributions and potential, the strength of the relationship and its potential to promote personal growth and scaffold the transition from childhood was evident in the photos, maps and dialogue with young people. For example, Quinn included photos of her support teachers from school at her school formal in her photo set, and talked about how important they had been to her (Figure 13.1). The role of one support worker with several young people who played and recorded music seemed

Figure 13.1 'Jamming together' by Luke

to provide a combination of practical support, motivation and a base of reassurance from which creativity welled.

Several young people either questioned or struggled with the level of authority they had over choice and decision-making in their lives to take up the opportunities they wanted. This related to both choices and decisions at a routine small-scale level and at broader levels about lifestyle and structural issues such as where to live or how to spend their week. During the research, the young people primarily talked about and made decisions about small-scale issues rather than larger ones, although some people referred to having made big decisions, such as buying a car, having surgery, and deciding which disability support service to use after they left school. A few people talked about how the days and number of days a week they participated in disability services or worked was determined by agencies, not by their preference.

Others struggled in accepting care that they did not want, particularly as they became young adults. One young person had been forcibly removed from their family into the child protection system. One participant talked about the difficulty negotiating with support workers who provided lacklustre care, resulting in repeated 'outings' to the shops and little personal development. Jums very eloquently raised an ongoing embodied tension about the difficulty receiving personal assistance as a young woman with disability, saying,

> this is a picture of how I guess I have to be supported. And we called it 'unwanted support' cause I'd rather be independent. So even though I need it, doesn't necessarily mean I want someone there (see Figure 13.2).

Figure 13.2 'Unwanted support' by Jums

Young people's care and caring

This research explored how young people with cognitive disability viewed care in their relationships by using participatory research methods that enabled them to express their perceptions through various media and activities. It found that young people spoke about or portrayed themselves as active agents in mutual care and peer relationships with friends and family. The mutuality of the caring relationships contributed to their personal identity, where they felt recognised, viewed as capable, respected and valued. Paid support services facilitated some relationships for people with higher support needs.

Applying Sullivan et al's (2016) framework of utility, agency and opportunity highlighted that understanding the quality and dynamics within the relationships from the perspective of the young person was as important as the type and number of relationships. They were concerned about whether they felt supported in their agency within the relationship, the functionality it performed and the opportunities it enabled. They expressed this in terms of mutual care that demonstrated the reciprocal valuing of the young person as caring and cared about. Interestingly for this book, they expressed this in language and meanings about their role in care and caring with intimate and more distant others. Their expression of the importance of this mutuality was in words, images and actions about when relationships felt good, were useful and enabled opportunities to enact being a young person in their community. Conversely, they felt devalued when that agency and opportunities for reciprocity of care was denied, whether benevolently, in the name of risk aversion or discriminatorily, due to their disability. As young people with cognitive disability in small communities they were particularly vulnerable to these negative aspects of relationships due to power differentials with other young people, adults and service providers.

Implications of these findings are that recognising the importance of agency in relationships and places is important to young people to develop and maintain a sense of identity and connection. Young people are embedded in a range of roles and relationships. They are active agents in these roles, both giving and receiving support and care in multiple ways. The reciprocity of care, especially in family life, was important to several young people in this study, who strongly identified with their emerging role as a provider of care within the family unit as an adult, as well as receiving support and care from parents. The tension between independence and reliance on others is a difficult one, particularly as young people leave childhood and negotiate multiple transitions. Some young people in this research experienced a struggle between agency and dependency. Having little choice over the provision of needed care made it harder for young people, especially those who need personal care and physical support. Having as much control as possible mattered.

The second implication is that relationships are critical for establishing and sustaining young people's sense of agency. When relationships support young people to thrive, they have shared qualities. They are mutual and reciprocal, and both partners recognise positive qualities and contributions in each other. They are underpinned by respect, and this is embodied and demonstrated in actions and interactions. They are evolving and responsive to change, and reinforce capability and protect against isolation and loneliness. They develop and strengthen over time. The findings about the high incidence of loneliness indicate that many young people with cognitive impairment would benefit from support to develop respectful friendships, not only for reasons of utility and opportunity, but also to enhance their agency through reciprocal care.

Far from opportunities for reciprocal care, the high incidence of violence and interpersonal harm had very negative impact on some of the young people. Interpersonal harm in particular tends to go 'under the radar' of policy frameworks, which focus on responding to incidents of abuse and neglect (Stalker and MacArthur 2012, Robinson and Chenoweth 2011). The nature of many of the harms experienced by young people in this research are different – they are pervasive, often multiple, and have a cumulative effect. The individual and collective impact of this harm needs further exploration with young people and those who support them.

This chapter has engaged in a broad sense of care in the lives of young people with cognitive disability. When they were invited to participate in research about people and places in their lives, their expressions about positive care were about reciprocity within their relationships. Their identity was as care givers as much as care receivers, emphasising their capacity and value. These encouraging findings imply clear direction for practice, policy and theory – enhancing the agency of young people through respect for their mutual contributions within relationships has immediate reward from a sense of worth, as well as utility to pursue the opportunities of any young people in their community.

Note

1 Cognitive disability is a term increasingly used in Australia to include a range of people who have impairments such as intellectual disability, acquired brain injury, specific learning disabilities, neurological conditions, and those on the autism spectrum. This article applies the current Australian preference for people-first language consistent with the UN Convention on the Rights of Persons with Disabilities.

References

Anderson, K., Balandin, S. and Clendon, S. (2011) 'He cares about me and I care about him': children's experiences of friendship with peers who use AAC. *Augmentative and Alternative Communication* 27, pp.77–90.

Australian Bureau of Statistics (2012) Year Book Australia 2012. Canberra, Australian Bureau of Statistics.

Day, C. (2015) Young adult carers: a literature review informing the re-conceptualisation of young adult caregiving in Australia. *Journal of Youth Studies* 18, pp.855–66.

Fitzgerald, R., Graham, A., Smith, A. and Taylor, N. (2010) Children's participation as a struggle over recognition. In Percy-Smith, B. and Thomas, N. (eds.) *Handbook of children's participation: perspectives from theory and practice*. London, Routledge, pp.293–305.

Hall, E. (2005) The entangled geographies of social exclusion/inclusion for people with learning disabilities. *Health and Place* 11, pp.107–15.

Hall, E. (2010) Spaces of social inclusion and belonging for people with intellectual disabilities, *Journal of Intellectual Disability Research* 54, pp.48–57.

Hamilton, M.G. and Adamson, E. (2013) Bounded agency in young carers' lifecourse-stage domains and transitions. *Journal of Youth Studies* 16, pp.101-117

Hill, M. (2002) Network assessments and diagrams: a flexible friend for social work practice and education. *Journal of Social Work* 2, pp.233–54.

Honneth, A. (1995) *The struggle for recognition: the moral grammar of social conflicts*. Cambridge, MIT Press.

Jahoda, A. and Pownall, J. (2014) Sexual understanding, sources of information and social networks; the reports of young people with intellectual disabilities and their non-disabled peers. *Journal of Intellectual Disability Research* 58, pp.430–41

Kayess, R. and French, P. (2008) Out of darkness into light? Introducing the Convention on the Rights of Persons with Disabilities. *Human Rights Law Review 8*, pp.1–34.

Nolan, M., Grant, G. and Keady, J. (1996) *Understanding family care*. Buckingham, Open University Press.

Power, A. (2008) Caring for independent lives: geographies of caring for young adults with intellectual disabilities. *Social Science and Medicine* 67, pp.834–43.

Power, A. (2013) Making space for belonging: critical reflections on the implementation of personalised adult social care under the veil of meaningful inclusion. *Social Science and Medicine* 88, pp.68–75.

Rapanaro, C., Bartu, A. and Lee, A.H. (2008) Perceived benefits and negative impact of challenges encountered in caring for young adults with intellectual disabilities in the transition to adulthood. *Journal of Applied Research in Intellectual Disabilities* 21, pp.34–47

Robinson, S. and Chenoweth, L. (2011) Preventing abuse in accommodation services: from procedural response to protective cultures. *Journal of Intellectual Disability 15*, pp.63–74.

Robinson, S., Hill, M., Fisher, K., Graham, A. and Valentine, K. (2014) *In the picture: understanding belonging and connection for young people with cognitive disability in regional communities through photo-rich research, final report*. Lismore, Centre for Children and Young People.

Schneider, J., Wedgewood, N., Llewellyn, G. and McConnell, D. (2006). Families challenged by and accommodating to the adolescent years. *Journal of Intellectual Disability Research* 50, pp.926–36

Smyth, C., Cass, B. and Hill, T. (2011). Children and young people as active agents in care-giving: agency and constraint. *Children and Youth Services Review* 33, pp.509–14.

Stalker, K., and McArthur, K. (2012). Child abuse, child protection and disabled children: a review of recent research. *Child Abuse Review* 21, pp.24–40.

Sullivan, F., Bowden, K., McKenzie, K. and Quayle, E. (2016) The close relationships of people with intellectual disabilities: a qualitative study. *Journal of Applied Research in Intellectual Disabilities* 29, pp.172–84.

Wodehouse, G. and McGill, P. (2009). Support for family carers of children and young people with developmental disabilities and challenging behaviour: what stops it being helpful? *Journal of Intellectual Disability Research* 53, pp.644–53.

14 Globalising child circulation

The care of children who are privately fostered across international borders

Karen Wells

Introduction

Child circulation is a strategy of social protection (Fonseca 2002, Leinaweaver 2007). To think about the movement of children in these terms is to acknowledge the material reasons for placing children in the care of another. The existing literature on child circulation has focused on the movement of children between households as a cultural system that is responsive to the precariousness generated by poverty. Occasionally there is a mention of more affective reasons for placing children in the care of another household. These address the desires of adults to be parents or the desires of parents to extend their social networks. In this sense the migration of children is not unlike the migration of women on marriage to their husbands' homes in patrilineal societies. Children, like brides, mobilise and thicken networks between more or less distant kin.

This chapter explores some of the reasons why children are sent to the UK and how they settle in their new families. Children come to Britain without their parents from many different countries. A high-profile case in which an Ivorian girl who had been privately fostered in London, Victoria Climbié, was killed by her 'carer' (Laming 2003) produced a perception in the UK that most transnationally privately fostered children are from West Africa. In fact this is not the case, and we found a very diverse range of nationalities in our research.

This chapter analyses the experience of private fostering through the lens of care, which has been surprisingly absent in research on child migration. It proposes that transnational private fostering should be understood as a stretching across distance of local practices of care. The chapter begins with a synopsis of the literature on the care/migration nexus and how this may be applied to child migration. It then discusses the importance of family crisis and the breakdown of care to child migration; the experiences of care of children in their new families; and, finally, the role of religious belief in explaining the decision to care for someone else's child.

The chapter draws on data from an extended project on private fostering (see Acknowledgement) completed in December 2014. Private fostering is a legal term in the UK which refers to the care of children by families other than close kin. Families who privately foster children are required by law to notify social services if the care extends for longer than one month. Although social services monitor families to satisfy themselves that there are no child protection concerns in the new home, these children are not 'Looked After' (a UK statutory term for children in Local Authority care) and their birth families retain Parental Responsibility (a legal status) for their children. The legislation is chiefly intended to regulate the care of British children of British parents who are living away from their natal home for extended periods of time. The legislation does not make adjustments for children who are in the UK and whose parents are living in another country and who may not have legal entitlement to settle in the UK. Indeed, guidance for social workers on private fostering advises that social workers should be aware of children's immigration status with a view to organising their removal from the UK. Furthermore, since private fostering generally comes to the attention of social services when families or schools notify them, official statistics of fostering status are likely to be significantly lower than the actual numbers of families who are looking after a child who they are not legally related to. According to Coram BAAF in 2015 1,560 children were reported to local authorities as being cared for and accommodated in private fostering arrangements at March 31, 2015. During the year ending March 31, 2015 a similar number of private fostering arrangements began (2,740) as ended (2,770). BAAF commissioned research in 2010 that indicated that as many 10,000 children were in a private fostering arrangement at some point during their childhood, suggesting a considerable gap between notifications and actual arrangements.

Children who do not have leave to remain in the UK are to some extent protected from the impacts of immigration status since the removal of the British government's reservation in respect of immigration to the UK's ratification of the United Nations Convention on the Rights of the Child (UNCRC). However, once the child turns 18 this protection is removed and the impacts of not having a legal right to remain in the UK will impact sharply on their transitions to adulthood, for example their capacity to get work, to apply to university, to register for benefits and, indeed, to socialise. In addition they may face the threat of deportation.

Children and their foster parents may well be unaware of their immigration status or not seek to regularise it. It is much easier for children (under 18 years old) to make a successful application for citizenship than it is for adults to do so. British immigration law is extremely complex and anyone advising a child on this issue should seek advice from a qualified immigration solicitor. The data set consisted of eight narrative life history interviews with fostered children, four interviews with Social Workers, a focus group with four Social Workers (one of whom was also interviewed singly), a database of Advice Line calls (n=55) from Social Services to Children and Families Across Borders (CFAB), and a database of private fostering referrals by Local Authorities to CFAB (n=20). The research took place over 18 months from mid-2013 to the end of 2014, and was preceded by a pilot project in 2011.

Literature synopsis

The intersection between migration and care has focused mostly on the transnational family, and children are surprisingly absent in this literature. It has mostly attended to the formation of transnational families through the migration of mothers and their continuing 'mothering at a distance' (Parrenas 2001) of their children (Akesson et al 2012, Bryceson and Vuorela 2002, Hondagneu-Sotelo and Avila 1997, Hoang and Yeoh 2012, Horton 2009, Madianou 2012, Pratt 2012). Transnational fostering raises the possibility of a different kind of 'caring at a distance' when children care for/about others who they have 'left behind', sometimes siblings and sometimes parents, or parents care for/about their children 'at a distance' although it is their children who have left.

Parvati Raghuram has critiqued the focus of the intersection between care and migration on the generalising of care, and its inattention to what care might mean in different geographic contexts. She argues that this elides important dimensions of care including that "[m]igration may be a result of the breakdown of conjugal relations and subsequent abandonment by the rest of the family" (Raghuram 2012, p.165 citing Raghuram 2005). This is an important insight, and the evidence that divorce and separation often precedes migration is scattered through the ethnographic accounts of transnational families. It is also recognised as a cause of the movement of children between households in domestic informal fostering. However, it has not yet been acknowledged in relation to the migration of children and young people across longer distances. The research discussed in this chapter found that child migration through transnational fostering is often, precisely, a result of the breakdown of family relations and abandonment.

Similar attention to the specific geographies of care is shown in Huang et al's interest in reciprocity of care between parents and (adult) children through the concept of filial piety in East Asian contexts. They suggest that this has expanded to include caring at a distance or indirect care (2012, p.131) of elderly parents by adult children who have now migrated. It may be that this stretching or de-territorialisation of care has now extended to parental care for (young) children, so that so long as resources (time, money, technology) are being deployed to enable a child to be cared for, this is also experienced by the child as being cared about, or at the least that the parents feel they are caring about their children. In our research, the children we interviewed did not feel that the deployment of resources to enable their migration was caring, and they felt rejected by their families. This is echoed in the few studies of children's feelings about mother's migration where children do not feel that sending money is compensation for losing the physical presence of their mothers (Horton 2009, Pratt 2012, Zentgraff and Chincilla 2012, p.353).

Transnationally fostered children are at the intersection of a number of policy regimes that are constantly shifting. The importance of the state in shaping migrant's experiences is often elided in the focus on individual stories, and on migrant's agency. Global care chain (GCC) analysis should avoid this kind of atomism and by definition take account of the different 'inputs' and 'outputs' in the care chain, including the state. Yeates excellent review of the global care chains literature establishes the explanatory power of global network analysis for attending to the "tangible mechanisms, structural outcomes and empirical impacts of care transnationalization" (2012, p.142). However, Raghuram maintains that insufficient attention is paid to the role of the state in delivering care in research on the migration/care nexus. This is a point also made by Kilkey and Merla (2014) who offer a structured typology of how policy regimes make it more or less difficult for care to be provided by/to migrants. Raghuram wants to see more integration between global care chains literature and the care diamond of market, state, community and family/household. She also questions the explanatory power of the care diamond, with the neat division between these four sectors implied by this metaphor. The study of child migration through transnational fostering necessitates attending to the intersection between these four sites of care because of the child's legal status as both a child and a migrant. This dual status, simultaneously the object of government care (towards children) and the object of government hostility (towards migrants) impacts on who can care for them and where (regulated by the state), the necessity of paying for their travel and papers, often through a smuggler/agent (organised through the market), the destination to a neighbourhood in which there is a relatively high number of families who are connected to the country of origin (emphasising the significance of community), and their departure from and attachment to a new family/household in the UK. This chapter explores primarily the last of these four points but this also touches on community, state and market – strengthening Raghuram's point that they are heuristic devices.

The migration/care nexus in GCC analysis only recognises children as the objects of care. This partly seems to echo the exclusion of children from the social reproduction literature, which GCC analysis parallels and in places intersects with. The GCC literature, like the social reproduction literature, acknowledges that children are at the heart of care and reproduction but leaves them as the objects being cared for/about/reproduced. Yet, I want to argue that transnational fostering and adoption are both part of global care chains. If global care chains are mobilised because of the uneven development of global capitalism which distributes the givers and recipients of care across the globe along trajectories that are often historically inscribed, then they are being activated whether the carer or the cared for is on the move. The transnational adopted child also might be considered as a migrant care worker, if, as David Eng (2006) has provocatively argued, the work that is expected of the 'migrant labourer' is not cooking and cleaning but the affective or emotional labour of completing the white middle class family in the heartlands of global capitalism. It is perhaps less provocative to argue that transnational fostering is a kind of care work, because the child is usually living with distant or social kin in these arrangements.

It is perhaps surprising that the literature on transnational adoption and transnational fostering has focused hardly at all on the concept of 'care'. The literature on transnational adoption has focused almost entirely on the parents' need to care for a child (to have a family) and the impact of 'elsewhere' on the child's identity

(Volkman 2005, Dorow 2006, Howell 2007, Yngvesson 2010) or dealing with it only to the extent that it shapes their cultural identity. In contrast, domestic adoption and fostering is framed around the best interests of the child. Of course, prospective adopters have their own investments in adoption but adoption agencies (private or public) conceptualise adoption and fostering as meeting the *child's* needs for care (both 'for' and 'about'). Indeed, it is this emphasis on the child's needs that allows agencies who accept payments for children to claim that they are not trafficking the child, although transnational adoption is increasingly being scrutinised as, at least potentially, a form of trafficking (Turner 2014). Transnational fostering is also under suspicion as a form of trafficking, particularly when foster families claim benefits or housing support for the foster child.

This also complicates the caring for/caring about typology widely used in the geographies of care literature (Milligan and Wiles 2010). In as much as care enters the picture at all it is 'care for' – meaning the material goods that the adoptive family can provide for the child in contrast to the poverty of the birth family. What the child might understand by being cared about is elided here with the presumption that being well cared 'for' can be directly translated to being well cared 'about', and conversely that it is impossible to a child to be cared about if they are not at the same time being cared for. Yet, for transnational fostering, if the foster parent attempts to get state support for the resources to care for their foster child, this places them under suspicion of trafficking because they are seen as benefiting financially from the movement of a child.

Transnational fostering seems to have largely been ignored by both migration scholars and in studies of kinship care. There are several studies by the government and NGOs of private fostering in the UK (MacLeod 2010, Morgan 2005, Owen et al 2007, Philpot, n.d., Shaw et al 2010, Utting n.d.) but few academic studies (Peart 2005). These studies tend to focus on the fostering of West African children who are assumed to be the majority of privately fostered children (Holman 1973, 2002, 2003, Longpet 2000, Olusanya and Hodes 2000). Our research suggests that this is not the case and that the countries of origin of transnationally fostered children are very diverse. Many children are from countries with colonial ties to the UK (e.g. Nigeria and Bangladesh) or from other European countries (especially Portugal and Spain) and the postcolonial states whose citizens have European citizenship through their histories, e.g. Macau/Portugal. Children who are transnationally fostered tend to be younger and are more likely to be girls than children who claim asylum; although both groups may arrive in the UK on false papers and having paid a smuggler, privately fostered children tend to arrive by safer routes (by air rather than overland). Kinship care, which is a child who lives with relatives other than their parents, is apparently increasing, with 175,000 children in kinship care according to the UK census, 95 per cent in informal arrangements. However, many of these children are living with grandparents and siblings, arrangements which do not fall within the scope of 'private fostering'.

One exception to the paucity of research on transnational fostering is the anthropology of child circulation in sub-Saharan Africa and in particular Esther Goody's seminal work on parenting, social reproduction and fostering in West Africa

(Goody 1982). Subsequent research on this practice in Africa has mostly focused on West Africa (Isiugo-Abanihe 1985, Bledsoe 2010 [1990], Castle 1995, Eloundou-Enyegue and Stokes 2002, Notermans 2003, Alber 2003, 2004) but there are now some studies on Central (Whitehouse 2009), East (Archambault 2010, Archambault and de Laat 2010) and South Africa (Madhavan 2004). Most of it is not transnational, focusing primarily on intra-country migration (but see Oien 2006). Nonetheless it provides some basis on which to understand historically prevalent and culturally embedded ideas about child fostering. There are multiple reasons why children move (or are moved) from one house to another. These movements are often temporary and maybe multiple, with a child moving to different households throughout his or her childhood (Archambault 2010). This reflects local ideas about what children need in their best interests. Among fostering families in Africa being attached to multiple households is seen as beneficial to children and their families. Generally, the point of circulation is that it is temporary, since it is this that creates a web of connection between the two or more households that the child has lived/is living in.

Underlying these 'migrations' and making them possible is a conception of children as being adaptive. The idea that the young child has a singular attachment to his or her birth mother seems to be largely absent in cultures where fosterage is practised on a wide scale. This does not mean that children are not valued. They clearly are and indeed social personhood is deeply tied up with child-rearing. For this reason infertility is a common reason for a child to be given to another household (Archambault 2010, p.233). Alber in her ethnography of Baatombu child circulation in Northern Benin notes that it is because children are highly valued that they are given away:

> as many people explained to me, if one gives one's child away, one gives the most valued thing one could give. Hence, the person who takes a child will be forever grateful for having received the most precious gift one could possibly receive. The great value of children makes the gift of a child particularly precious. It is this great value attributed to children that qualifies them as ideal objects of exchange.
>
> (Alber 2003, p.499)

Alber's use of the term 'objects of exchange' in the above quote brings into focus the status of the child as belonging to parents; to make a 'gift' of a child implies both reciprocity (it will be given back) and (inalienable) ownership. The failure to return the gift obligates the recipient (see Yngvesson 2002). This should not be understood as a direct, acknowledged obligation of reciprocity since it is precisely the character of a gift that it obscures the obligations inscribed on/in it. At the very least, however, it confirms the sense that the gift of the child creates social networks or webs of relation between giver and recipient. It also suggests the prevalent view in non-capitalist cultures of children as 'gift' rather than 'cost' or 'burden'. This contrasts, for example, with the underlying premise of much social policy in the Global North that considers children as human capital, and therefore as a cost that should only be borne if the anticipated financial returns are sufficient (Green and Lawson 2011). This underlies, for example, social policy that is hostile to teenage parents, to poor and working class people having large families and persistent attempts to calculate the financial cost of raising children. This different worldview on children goes some way to explaining why families will care for (informally fostered) children for whom they have no legal or even familial obligations.

Transnational fostering or child circulation raises interesting questions about what it means to care for or about a child and indeed, how children care for and about each other and their families. These two kinds of movement of children (that is within a city or country or between countries) from their natal families are not usually considered together, and this in itself raises interesting questions about how we theorise children's lives. More generally, because lone child migration forms part of a complex of child migration in which the child is primarily constituted as a problem of governance, the *care* of child migrants has attracted very little attention. To focus on the place of care in understanding child migration this chapter has reviewed the literature on the migration/care nexus and on transnational adoption and fosterage. In the analysis of the migration/care nexus, children have been seen as the 'objects' that mobilise the migration of adults. However, the fact that children move to be cared for, rather than adults moving to care for them, does not change the fundamental characteristic of a care chain: "personal links between people across the globe based on the paid or unpaid work of caring" (Yeates 2012, p.145). GCC analysis is based on the idea that a globalised system of production demands, or at least will generate, a global system of (social) reproduction. That is, that the stretching of relations of material production will be accompanied by the stretching of relations of social reproduction. For the most part, that has not happened; most social reproduction, especially involving children, gets done very locally. Nonetheless there is to some degree, and perhaps increasingly, a stretching of the relations of social reproduction and the formation of a transnational field. Raghuram notes that the assumption that women migrate and leave their children behind in a calculation of economic costs and benefits hides the fact that migration is often also propelled by crisis.

Family crisis

The centrality of family crisis to the transnational fostering of the children in our study was a surprising and significant finding. Family crisis, usually involving the death of a parent, parental separation or parental abandonment was a feature of all of the children's life history narratives and also came through strongly in the referrals and advice line. Interestingly, this was not at the forefront on the social workers accounts of child migration. Similarly, in the wider literature on independent child migration this has not been highlighted. We think this is an important finding, and one that is specifically related to *child* migration and the importance of care in the lives of children, in contrast to other forms of child migration (e.g. asylum and forced migration) which share many similarities with adult migration.

In a review of the literature on HIV/AIDS, migration and family care, Haour-Knipe (2009) found that parents' illnesses and death were often a cause of child migration. Ansell and van Blerk (2004) also found that parental sickness and death were a cause of child migration, with significant numbers of child respondents in Malawi and Lesotho having migrated following their parents' death or sickness and as many as one in five in Lesotho and one in three in Malawi now living with children who had come into their households following parental death or sickness. In her research on the street-ward migration of young Ugandans, Lorriane Young (2004) found that a parent's death was cited by one in five children as a reason for leaving home. However, she also cautions against over-simplifying the impact of family crisis on child migration and recognising the importance of "[t]he spatial integration of structural conditions at the national level with family crisis at the local level". Although most of the (still scant) literature on child migration and family crisis is based on research into the impact of HIV/AIDS on families/households in sub-Saharan Africa, whether parental sickness and death is caused by HIV/AIDS or some other reason, the underlying logic of a coping strategy based on migration remains.

This research echoes their findings: parental abandonment, divorce and parental death were important events structuring children's migration narratives, both in interviews and in advice line and referrals to CFAB. One boy, Abbas (all names are pseudonyms) told me that he came to the UK because his mother had died when he was two or three years old. His father remarried after two years and when Abbas was seven he was sent to the UK. His father now lives in Kuwait and his stepmother and her children stayed in his country of birth. When he came to the UK he was initially left with a family but those people contacted a single woman who had no children and she took him in. Another boy who is now living with a family in London says that his mum left him when he was very young and "my dad never came back". He says he was cared for by strangers (the servants who had lived in the house before his parents left) until they brought him the UK when he was 11. Although this boy's story had elements that may seem fictitious or unbelievable, the general gist that he was an only child whose parents left him (whether in his birth country as he claims, or in London as seems more likely) the central contention that a crisis precipitated his movement to the UK is nonetheless believable.

The only girl in our life history narrative sample was a (now) 15-year-old who had been sent to London with an agent paid for by her grandfather. Her mother had left her and her brother when her brother was a little boy. She had then lived with her grandparents and her brother until her grandmother passed several years earlier. Her grandfather remarried so that there would be someone to care for his grandchildren, but the 'step-nanny', as she called her, was not kind to the children. In addition to the crisis in care precipitated first by her father's death, then her mother's abandonment and then her grandmother's death, Dukah was then subjected to sexual harassment and assault by a neighbour. This man was charged and imprisoned for some time but on his release, the grandfather.

> was, like, annoyed with me: "look, your mum ran away, now because of you, you are bringing shame"... he planned to get rid of me, then he found out some way to send me, because here, one of my aunt lives.... So, he says, okay, if I just send you another country, then I will be, like, I will be in peace without you.

Dukah's lengthy and detailed account attests to the importance of networks in seeking a solution to a familial crisis, in this case a series of losses and abandonments, compounded by shame and the step-grandmother's harshness.

Although, as we mentioned, social workers seem rather uncertain if family crisis was a trigger for migration, they also described several cases where this was the case. In one case, the mother was arrested for drug smuggling on entry to the UK and in another case the mother was very young and sent her child to the UK from Jamaica to be looked after by her grandmother's sister. In two cases, children were in private fostering after one or both parents died in a war. In the referrals to CFAB a quarter (5/20) involved some kind of family crisis including parental abandonment, maternal death and conflict with step-parents. In the audit of advice line calls, 6/55 involved family crisis including parental abandonment, parents in prison or parental death. There is some evidence of a pattern of family crisis which involves father's death followed by maternal abandonment.

To conclude this section, in a significant number of cases and responses family crises was a key theme in explaining child migration to the UK. This finding is confirmed by research by Sarkar et al (2003) into informal fostering in the slums of Dhaka. They noted both that informal fostering was widely practised by the poor and that maternal death accounted for nearly two-thirds (65 per cent) of the reasons for fostering. Yet, this is largely absent in the wider literature on lone child migration which tends to assume (e.g. in relation to young asylum seekers and refugees) that migration is essentially structural and either strategic or forced.

New families

The children in this study were all known to social services (because we recruited our sample through social services and social services contacted CFAB). Most had been referred to social services through the United Kingdom Border Authority (UKBA) when their carers tried to regularise the child's immigration status or by the child's schools. The children involved in our research were remarkably resilient and adaptable and spoke of their intimacy with their 'cousins' and affection for and attachment to their carers. In this section we describe their accounts of new family life focusing on three cases in which the children elaborated most on their family life. Although we focus on these three respondents, the other respondents were also positive about their new families: "They always respect me, they always play with me, they're always talking to me," said Bakri. Explaining how he had come to adapt to living in London with his new family Bakri said, simply, "[I wanted my parents to] come back, come back, but they didn't come back, then I find [went to] my little brother and I play with them, and I stay with them. They stay with me and I was with them." The social workers also said that "some of the carers are fantastic. Some of the private foster carers are really fantastic carers."

Dukah has also settled well with her family in London. The start of living in London was difficult. Her grandfather had not told her that she was moving to London but rather he had suggested that she should "go away, have a little visit different country, then maybe everything will cool down, you will feel better". So, when she arrived in London she thought it would be for a holiday. When her aunt, who her grandfather had sent her to, realised his intention she went to the girl's grandfather to persuade him to take the girl back but he refused. The aunt had left Dukah in the care of an uncle while she went to visit the grandfather. When she failed to persuade the grandfather to take Dukah back the aunt returned to London and told the 'uncle' (actually her father's cousin),

> My family is not going to accept her. They are saying no, we've got our own privacy in our house; we're not going to let her. She can come, yes, she is my cousin, as a visit, she can visit, fine, but not permanently … and she was like, oh my days, now, I'm in a great trouble; what I'm going to do with this girl? Then my uncle says, okay, I'm going to take care of her.

Dukah now lives with her uncle and aunt and her three cousins. Her youngest cousin is five and the other three are older than her. She shares a bedroom with her girl cousin, who is 20 and who she calls cousin-sister. She says,

> our relation is, like, so close, we, I don't feel she's older than me or something. I mean, we are so close and she is really friendly and she helps me with everything if I need any help with my school work or in home; if I need anything, I ask her, I ask my aunt. Yes, everything's going on fine.

She has only been living with this family for 18 months. Although she knew them when she was much younger she had not seen her uncle or her cousins for seven or eight years. Unlike the other children in our sample she has a sibling in her birth country, a younger brother, who she keeps in touch with. He still lives with her grandfather and step-nanny and she is concerned for his welfare.

Children making families

> I live with the carer and she's really caring; because she hasn't got no kid's she's like, she look after me like her own kids.
>
> (Abbas)

Research on transnational adoption, as I mentioned earlier, emphasises the importance of children in the making of families. Similarly, research on fosterage in

West Africa explains childlessness as one reason that a child might be sent to live in another household. We can suggest that caring for and about a child is one way to make a family. Research in Bangladesh in 1998 on informal fostering with 46 foster mothers caring for foster children under 24 months found that more than half of the foster mothers were childless (Sarkah et al 2003). It is interesting that two of the children in our small sample of respondents were fostered by childless women.

Abbas came to the UK six or seven years ago and is now 16 years old. He lives with his carer in a one-bedroom flat. She does not have any children and Abbas said that although she is married she does not live with her husband. Abbas was one of two children in our sample who lived with a single woman. In a sense these children make the women who care for them part of a family. Abbas says he is also the only boy in an extended family of girls. His foster mum (as he calls her) has a sister-in-law and a brother-in-law and nieces but no nephews. He implied that this was a significant explanation of why they were so pleased to have him in the family:

> [when] I first came he [his foster mum's the brother-in-law] brought me, that time was my birthday, he bought me games, went shopping and bought me everything because their whole family they haven't got no sons, all daughters, like, no sons.

The other boy who lived with a single woman, Najid, also shares a one-bedroom flat with his carer who he calls auntie. She has applied for a two-bedroom accommodation for her and the boy but has been refused. He is perhaps less attached to his auntie than Abbas is to his 'mum' and in response to my question "Do you think she loves you?" he replied "Yes, I think so, something, yes". Yet he does seem very attached to her. He has lived with her now for six years (since he was eight); they share a bedroom which, he says is, "a bit difficult for us, but still". It is because of her, he says, that he "got more into religion" and his religious belief and practice gives him a lot of comfort.

Second family crisis

Although most of the children settled well with their new families the social workers also identified cases where family life in London was not so positive. One child is now looked after under Section 20 having been physically abused by his carers, and two children in another family were working as (unpaid) domestic servants, while the foster carers' own children of a similar age went to school. Another boy had been sexually abused when his initial private fostering arrangement broke down and the 'carers' found someone on the market for the boy to live with. The boy disclosed sexual abuse to his school a couple of days later and was taken into Local Authority care. Less extreme but still difficult was another case of a child living with a single woman who was finding it difficult to look after the boy now and wanted him to be taken into care although the boy wanted to carry on living with his carer. Another child was taken into care because the carer contacted social services and said he would no longer provide financially for the child and social services should come and pick him up. The majority of cases referred to CFAB involved a breakdown of the first care arrangements that had been made informally between families. These cases highlight the importance of the state as part of the migration/care nexus that we referred to in the synopsis of the literature.

Islam: caring for others and the care of the self

Katy Gardner in her article on 'the transnational work of kinship and caring' says that being a 'good woman', to her Bengali respondents, is "inseparable from being an unquestioning provider of care" which "in turn, is related to religiosity and one's relationship with God". As one of her respondents put it, "First you bow your head to the ground to Allah and then you do it for your husband" (Gardner, 2006, p.383). Gardner and Ahmed develop these points in their paper on informal social protection in which they discuss the duty of care that Bengali Muslims have to their 'own poor'. In addition to the general obligation of Muslims to give alms (one of the five pillars of Islam), "obligations to provide help increase according to the closeness of the connection … degrees of relatedness are constructed over time and calculated according to where it is that people originate from" (Gardner and Ahmed 2009, p.137).

Other than Gardner's work there is a dearth of academic literature that explores the relationship between Islamic ethics, diasporic identity and child circulation. The Prophet himself was an orphan and there are several hadith that refer to the duty of Muslims to care for orphans. For example consider the hadith from Sahl ibn Sa'd: "The Prophet, peace and blessings be upon him, said, 'The one who cares for an orphan and myself will be together in Paradise like this', and he held his two fingers together to illustrate. Others include "The best house among the Muslims is one where an orphan is well treated, and the worst house among the Muslims is one where an orphan is badly treated" and "Would you like that your heart becomes soft and that you acquire what you need? Be merciful with the Orphans, pat his head and feed him from what you eat."

These hadith do not in themselves make the care of parentless children an Islamic duty but they do provide, perhaps, a religious-cultural context within which the decision to accept an 'orphan' into one's family makes sense.

The concept of a global Islamic community, further strengthened by the importance of the concept of *desh* to Bangladeshis in London is important in explaining these children's experiences and the moral/religious logic of those that take them in. The social workers also affirmed this sense of religious obligation on the part of some foster carers. I asked Dukha why she thought her 'far relative' would take her in when her "dad's own blood … don't even care about us. But he is looking after me; he's doing everything for me":

> He says, because I'm helpless. Because I am helpless; it's something about humanity. He just, he is just helping me, because in our religion, if you help an orphan, there's lots of good things. Yes, because he is a really religious person and he's helping me because I'm helpless. I have nowhere to go. In Bangladesh, I'm not safe either; I have nowhere to go. Where would I go?

Conclusion

The care needs and care practices of children migrating alone have been surprisingly absent in both academic research and policy debates. Literature on the care/migration nexus has primarily focused on the experience of the migrating parent and their attempts to mother at a distance. The presumption that migration is a rational choice in which the parent chooses one option among many for the child's future is not supported by our findings. These suggest instead a rather chaotic, making-do approach to migration in which the need to find someone to care for a child who can no longer be cared for at home can be potentially met by people thousands of miles away. Private or informal fostering on a transnational scale is a stretching of long-standing practices of tapping into social networks, but these networks now extend across the globe. It is perhaps because they are historically and culturally embedded practices of care that for the most part the children are accepted into their new families so quickly and seem to form close bonds with their carers and cousins in a short space of time. Islam plays a role here in explaining to children why families take them in, as well as in the decisions of parents to take on the care of (another) child and perhaps also in the adaptability of the children to culturally familiar homes albeit in a strange land. In many families, children share bedrooms and of course food and their parent's time and attention with the new child in the family, but in a care nexus where children are seen as a gift rather than a burden, the resources they need are not begrudged by either the adults or their new cousin-brothers and cousin-sisters. It is however important to note that the life history narratives that this chapter draws on in its analysis are from children who have agreed to participate in a study and who are known to social services; they are not necessarily representative of the general experience of care of transnationally fostered children. However, what their experience can tell us is that child migration does not, in itself, lead to a deficit in care and indeed may lead to a more caring environment for children if their natal home is in crisis, particularly following parental death.

Care is an interesting lens through which to view children's experience of movement and settlement, of migration and immigration. It re-focuses our analysis on the child as a living person with a complex subjectivity rather than an object of social policy. Thinking of transnational movement as a movement towards care also unravels the borders between nation-states. It opens up a mapping of the world inscribed by web of social connections, often historically embedded and always culturally formed, in place of the more recent view of the world as a division of nation-states or regional blocs.

Acknowledgement

This research was conducted by the author, Birkbeck, University of London, in collaboration with the UK-based NGO, Children and Families Across Borders and partly funded by both institutions and the British Association for the Study and

Prevention of Child Abuse and Neglect (BASPCAN), whose support is gratefully acknowledged. A pilot study was conducted in 2011; the data analysed for this paper was mainly collected in 2014.

References

Åkesson, L., Carling, J. and Drotbohm, H. (2012) Mobility, moralities and motherhood: navigating the contingencies of Cape Verdean lives. *Journal of Ethnic and Migration Studies* 38, pp.237–60.

Alber, E. (2003) Denying biological parenthood: Fosterage in Northern Benin. Ethnos 68(4), pp.487–506.

Alber, E. (2004) Grandparents as foster-parents: transformations in foster relations between grandparents and grandchildren in northern Benin. *Africa* 74, pp.28–46.

Ansell, N. and van Blerk, L. (2004) Children's migration as a household/family strategy: coping with AIDS in Lesotho and Malawi. *Journal of Southern African Studies* 30, pp.673–90.

Archambault C. (2010), Fixing families of mobile children: Recreating kinship and belonging among Maasai adoptees in Kenya. *Childhood*, 17(2), pp.229–242.

Archambault, C. and de Laat, J. (2010) Social mobility in children's mobility? An investigation into child circulation among the Maasai of Kenya. In Ensor, M. and Goździak, E. (eds.) *Children and migration: At the crossroads of resiliency and vulnerability.* Basingstoke, Palgrave Macmillan pp.187–206.

Bledsoe, C. (1990) 'No success without struggle': social mobility and hardship for foster children in Sierra Leone. *Man* 25, pp.70–88.

Bryceson, D. and Vuorela, V. (eds.) (2002) *The transnational family: new European frontiers and global networks*. Oxford, Berg.

Castle, S. (1995) Child fostering and children's nutritional outcomes in rural Mali: the role of female status in directing child transfers. *Social Science and Medicine* 40, pp.679–93.

Dorow, S. (2006) *Transnational adoption: a cultural economy of race, gender, and kinship*. New York, New York University Press.

Eloundou-Enyegue, P. and Shannon Stokes C. (2002) Will economic crises in Africa weaken rural-urban ties? Insights from child fosterage trends in Cameroon. *Rural Sociology* 67, pp.278–98.

Eng, D. (2006) Political economics of passion: transnational adoption and global woman. *Studies in Gender and Sexuality* 7, pp.49–59.

Fonseca, C. (2002) The politics of adoption: child rights in the Brazilian setting. *Law and Policy* 24, pp.199–227.

Gardner, K. (2006) The transnational work of kinship and caring: Bengali British marriages in historical perspective. *Global Networks*, 6(4), pp.373–389.

Gardner, K. and Ahmed, Z. (2009) Degrees of separation: informal social protection, relatedness and migration in Biswanath, Bangladesh. *Journal of Development Studies*, 45(1). pp.124–149.

Goody, E. (1982) *Parenthood and social reproduction: fostering and occupational roles in West Africa*. Cambridge, Cambridge University Press.

Green, M. and Lawson, V. (2011) Recentring care: interrogating the commodification of care. *Social and Cultural Geography* 12, pp.639–54.

Haour-Knipe, M. (2009) Families, children, migration and AIDS. *AIDS Care* 21, pp.43–8.

Hoang, L. and Yeoh, B. (2012) *Sustaining families across transnational spaces: Vietnamese migrant parents and their left-behind children*. Asian Studies Review 36, pp.307–25.

Holman, R. (1973) *Trading in children: a study in private fostering*. London, Routledge.

Holman, R. (2002) *The unknown fostering: a study of private fostering*. Lyme Regis, Russell House.

Holman, R. (2003) Private fostering: old problems, new urgency. *Adoption and Fostering* 27, pp.8–18.

Hondagneu-Sotelo, P. and Avila, E. (1997) 'I'm here but I'm there': the meanings of Latina transnational motherhood. *Gender and Society* 11, pp.548–71.

Horton, S. (2009). A mother's heart is weighed down with stones: a phenomenological approach to the experience of transnational motherhood. *Culture, Medicine and Psychiatry* 33, pp.21–40.

Howell, S. (2007) *The kinning of foreigners: transnational adoption in a global perspective.* New York, Berghahn.

Huang, S., Thang, L. and Toyota, M. (2012) Transnational mobilities for care: rethinking the dynamics of care in Asia. *Global Networks* 12, pp.129–34.

Isiugo-Abanihe, U. (1985) Child fosterage in West Africa. *Population and Development Review* 11, pp.53–73.

Kilkey, M., and Merla, L. (2014) Situating transnational families' care-giving arrangements: the role of institutional contexts. *Global Networks* 14, pp.210–29.

Laming, Lord W.H. (2003) The Victoria Climbie Inquiry. Report of an Inquiry by Lord Laming. London,The Stationary Office.

Leinaweaver, J. (2007) On moving children: the social implication of Andean child circulation. *American Ethnologist* 34, pp.163–80.

Longpet, H. (2000) *Private fostering of children of West African origin in England.* Unpublished PhD thesis, University of Bristol.

MacLeod, M. (2010) *No simple answers: report of the DCSF advisory group on private fostering 2008–10.* London, Department for Children, Schools and Families.

Madhavan, S. (2004) Fosterage patterns in the age of AIDS: continuity and change. *Social Science and Medicine* 58, pp.1443–54.

Madianou, M. (2012) Migration and the accentuated ambivalence of motherhood: the role of ICTs in Filipino transnational families. *Global Networks* 12, pp.277–95.

Milligan, C. and Wiles, J. (2010) Landscapes of care. *Progress in Human Geography* 34, pp.736–54.

Morgan, R. (2005) *Private fostering: some views from privately fostered children on the government's proposals about private fostering*. London, Commission for Social Care Inspection.

Notermans, C. (2003) Nomads in kinship: fosterage and the self in Cameroon. *Focaal* 42, pp.89–103.

Øien, C. (2006) Transnational networks of care: Angolan children in fosterage in Portugal. *Ethnic and Racial Studies* 29, pp.1104–17.

Olusanya, B. and Hodes, D. (2000) West African children in private foster care in City and Hackney. *Child Care, Health and Development* 26, pp.337–42.

Owen, C., Jackson, S., Barreau, S. and Peart, E. (2007) *An exploratory study of private fostering*. London, TCRU.

Parrenas, R. (2001) Mothering from a distance: emotions, gender, and intergenerational relations in Filipino transnational families. *Feminist Studies* 27, pp.361–90.

Peart, E. (2005) The experience of being privately fostered. *Adoption and Fostering* 29, pp.57–67.

Philpot, T. (n.d.) *A very private practice: an investigation into private fostering.* London, BAAF.

Pratt, G. (2012) *Families apart: migrant mothers and the conflicts of labor and love.* Minneapolis, University of Minnesota Press.

Raghuram, P. (2005) Global maid trade: Indian domestic workers in the global market. In S. Huang, N. Abdul Rahman, and B. Yeoh (eds.) Asian transnational domestic workers. Singapore, Marshall Cavendish, pp.146–74.

Raghuram, P. (2012) Global care, local configurations – challenges to conceptualizations of care. *Global Networks* 12, pp.155–74.

Sarkar, N., Biswas, K., Khatun, U. and Datta A. (2003) Characteristics of young foster children in the urban slums of Bangladesh. *Acta Paediatrica* 92, pp.839–42.
Shaw, C., Brodie, I., Ellis, A., Graham, B., Mainey, A., de Sousa, S. and Willmot, N. (2010) *Research into private fostering*. London, Department for Children, Schools and Families.
Turner, C. (2014) Sometimes it is better not to be unique: the U.S. Department of State view on intercountry adoption and child trafficking and why it should change. *Duke Forum for Law and Social Change* 6, pp.91–112.
Utting, W. (n.d.) *People like us: the report of the review of the safeguards for children living away from home*. London, HMSO.
Volkman, T. (ed.) (2005) *Cultures of transnational adoption*. Durham NC, Duke University Press.
Whitehouse, B. (2009) Transnational childrearing and the preservation of transnational identity in Brazzaville, Congo. *Global Networks* 9(1), pp.82–99.
Yeates, N. (2012) Global care chains: a state-of-the-art review and future directions in care transnationalization research. *Global Networks* 12, pp.135–54.
Yngvesson, B. (2002) Placing the 'Gift Child' in transnational adoption. *Law and Society Review*, 36(2), pp.227–255.
Yngvesson, B. (2010) *Belonging in an adopted world: race, identity, and transnational adoption*. Chicago, University of Chicago Press.
Young, L. (2004) Journeys to the street: the complex migration geographies of Ugandan street children. *Geoforum* 35, pp.471–88.
Zentgraf, K. and Stoltz Chinchilla, N. (2012) Transnational family separation: a framework for analysis. *Journal of Ethnic and Migration Studies* 38, pp.345–66.

15 Reflection 1

Distance, connection and the power, freedom and obligation (not) to care

Kathrin Hörschelmann

Amongst the many different connotations carried by the word 'care', connectedness is perhaps one of the most central. Care is often assumed to be based on a connectedness that is deemed to exist naturally and/or to come from 'the heart', i.e. from deep-seated emotions and senses of closeness. Yet, as the texts in this collection make clear, such connectedness and the 'desire' to care rarely exist automatically or freely. As relational, social beings, our senses of who or what to care for are strongly bound up with social expectations and with our complex, shifting positions within social relations, including relations of gender and generation. We often notice this most when our sense of care, and perhaps the feeling that we do not wish to care or are not cared for, conflict with weighty norms and societal expectations. Indeed, the demand to care is often a way to put us in our place and it causes significant tensions between developing positively connoted subjectivities as caring individuals and refusing to care in the ways that others may demand.

This is not to reiterate the pessimistic, undermining and hurtful charge that care for others is motivated, and thus disqualified, by desires, needs and perhaps fears of the carer more than the needs of those cared for. It is to pay attention to the contradictions and ambiguities that many sense already when they (see themselves as people who) care about someone/something. Caring can lend us a positive sense of self, for sure, not least because we seem to willingly carry the costs it generally entails. It is also a way to express recognition of our relational entanglement with the lives of others and of our dependence on others and other things, not just close by, but also at a distance. In caring for others, we recognise ourselves as partial, while experiencing a degree of self-efficacy and perhaps even resilience in difficult circumstances. In caring, we also, however, most often have to bridge the gap of uncertainty, of not knowing whether and what kind of care is helpful or desired, a gap that post-structural theorists argue is an insurmountable characteristic of relationships.

Instead of simply turning the scepticism about motivations for care into its reverse, I would propose to engage more with the challenges presented by the authors in this book. One of the most important of these challenges, for me, is the need to consider care as an inevitably discriminatory practice, and to attend to the effects of this discrimination as well as to the ambiguities and contradictions that arise from it, expressed for instance in the exasperated exclamation, "But I cannot care for everyone/everything!"

How one cares, for whom, when and for how long are matters of negotiation and oftentimes of severe conflict, expressed across the spectrum of micro- and macro-political practices, whether in families, friendship groups, intimate relations and localised communal networks or through nationalism, the geopolitics of vulnerability (see Martin 2011), 'humanitarian' militarism or struggles over providing or refusing care for displaced persons and/or for non-human beings and things. And it is because of the ambiguities inherent in care, as well as, I would propose, the fact that we mostly encounter, and struggle over, absent/needed/denied/desired/given care in intimate relations, first and foremost, that makes this such a potent issue, so pliable for political purposes. The intimacy implied by care, most often, makes it appear 'natural' that 'we' *ought* to care most for those closest to us. This logic underpins not only the persistence of abusive adult–child relationships, but also nationalistic and racist politics. At the same time, those who wish for a different politics, one based on propinquity and senses of connectivity, often also see the need to promote an affective politics of closeness and of (inter)dependency, obliging 'us' to care. As de Leeuw shows in Chapter 4, such care is not 'innocent' per se, as it may also be based on ethnic discriminations and a colonial politics that so often is expressed in problematic terms of parent–child relationships and in normative understandings of care.

For research that seeks to untangle the complicated relations through which care, and caring/cared for subjectivities, are produced and performed intimately, institutionally and/or across distance, what seems crucial to me is thus to listen carefully, first and foremost, to different articulations and valuations of care (as done by virtually all of the authors in this volume), to consider both its benefits and costs as they are negotiated within care relations (cf. Evans Chapter 10 and 2011), to be attentive to the manifold ways in which those who are culturally constructed as cared for do themselves (have to/wish to) care (Aldridge and Becker 1993, Robson 2004), and to develop better understandings of reasons for *not* caring. Asking that latter question does not compel researchers to legitimise and valorise the denial of care, but impresses on us the need to explore the many purposes, moral enrolments and cultural connotations of care, including its use as a tool of oppression. As a former child enrolled multiply into more and less desired forms of care (for/by parents, siblings, socialist 'brothers and sisters', the state) and, today, as a parent and a privileged white European, writing this from the relative safety and material security of a well-paid elite worker, on *this* side of Europe's reinforced, lethal defence-system against caring for others, I think it would be hypocritical and dishonest not to at least acknowledge the underside of care that is also addressed by many of the authors in this volume.

Care appeals to seemingly heart-felt desires to extend beyond oneself and to share one's resources with those who (appear to) need them (more). Yet, this desire risks establishing relations of power on the basis of the *carer's* sense of somebody/something else in need of care. Relations of care establish obligations, as many of the authors in this collection show with reference to quite different contexts. On the other hand, care itself can become an obligation, something that is demanded of certain individuals *as if* it was a natural duty, something they *ought* to do. We become most aware of this sense of duty and obligation, of care as a normative demand that can be costly for the carer, when other social relations underpinning 'care' are seemingly turned upside down, when children care for other children, for adult family members and/or others in communal networks and beyond, as illustrated so vividly in this book.

The question of children as carers, and of the care responsibilities of adults, brings many of these issues to the fore. It shows how strongly care is tied to very specific cultural understandings and socially constructed identities. It is because of this that I think the particular relations of care by, between and for children, youth and adults, across different global contexts and as part of a politics of propinquity and connectivity (Amin 2004) deserves much greater attention than it has up to now, this volume and several notable exceptions not withstanding (cf. Aldridge and Becker 1993, Barry 2011, van Blerk and Ansell 2007, Evans 2011, Evans and Becker 2009, Robson 2004). Too often, we do not ask about the work that children and youth do within intimate and distanced relations, because it challenges adult subjectivities and brings many of the ambiguities I have tried to sketch above to the fore. Without recognising this work, in contexts of care, and the complicated circumstances and forms of care through which children co-produce and are enrolled in relations with various (non)caring individuals and institutions, the emotional benefits and costs of such care remain poorly understood and research risks contributing to (geo)political responses that are insensitive to the very needs they seem in turn to care for. Listening to articulated specific needs, acknowledging that care is often mutual and tied to expectations of reciprocity (cf. Hughes and Day, Chapter 2, Day, Chapter 9, Tronto 1993), and paying attention to the power relations and politics of emotion expressed within accounts of care and need is one of the 'duties of care' that we have as researchers, in my view. In order to do so, however, we need to be ready to challenge our own normative understandings of care and responsibility, and to be wary of naturalised assumptions about who cares, why, for whom. We also need to be prepared to listen to accounts by/of parents and other adults who do not care, who care differently, or who abuse children.

This is the point at which I would admit that deconstruction on the one hand and celebrations of young people's resilience and agency on the other hand take me to a place I hesitate to go, where I become anxious and unsettled. For the vulnerability of infants and the relative, if variable and contingent, dependencies of many children within intimate family, household, communal and/or institutional contexts are two aspects of the social that I find difficult to 'problematize' as social constructions (also see Wells 2009 on this issue). Sure, if by vulnerability and dependency we mean a lack of agency, sensitivity and capability, then they are problematic and misleading terms. However, recognition of emotional, psychological, bodily vulnerabilities strikes me as still crucial for calling up adult responsibilities towards children and for tackling injustice and abuse, both towards children one is close to and those whose lives are more or less directly entangled with one's own across different scales. It has taken a long time for child abuse to be given serious attention in sociological and psychological research and we have as yet a long way to go, I believe, to address the many failures of care that children are exposed to by virtue of generational power relations (amongst others) and their embodied vulnerabilities – which are all too often covered up by gendered myths of parental care. If we approach the issue of care from the perspective of child abuse, then the myths of motherhood, fatherhood and adult care for children that underpin so many of our general understandings of care need to be exposed to far more critical scrutiny and it may be that, as a result, new ways of caring with, for and between children, within a much wider range of contexts, distances and scales, will come to be envisioned.

References

Aldridge, J. and Becker, S. (1993) *Children who care: inside the world of young carers*. Loughborough, Loughborough University.

Amin, A. (2004) Regions unbound: Towards a new politics of place. *Geografiska Annaler* 86(1), pp. 33–44.

Barry, M. (2011) 'I realised that I wasn't alone': the views and experiences of young carers from a social capital perspective. *Journal of Youth Studies* 15, pp.523–39.

Evans, R. (2011)'We are managing our own lives…': life transitions and care in sibling-headed households affected by AIDS in Tanzania and Uganda. *Area* 43, pp.384–96.

Evans, R. and Becker, S. (2009) *Children caring for parents with HIV and AIDS: global issues and policy responses*. Bristol, Policy Press.

Martin, L. (2011) The geopolitics of vulnerability: children's legal subjectivity, immigrant family detention and US immigration law and enforcement policy. *Gender, Place and Culture* 18(4), pp.47–498.

Robson, E. (2004) 'Hidden child workers: young carers in Zimbabwe', *Antipode* 36, pp.227–48.

Tronto, J. (1993) *Moral boundaries: a political argument for an ethic of care*. New York, Routledge.

van Blerk, L. and Ansell, N. (2007) Alternative care giving in the context of AIDS in southern Africa: complex strategies for care. *Journal of International Development* 19, pp.865–84.

Wells, K. (2009) *Childhood in a global perspective*. Cambridge, Polity Press.

16 Reflection 2

Providing care: challenges for practitioners and service providers

Michelle Pyer and John Horton

Working with definitions of …

One of the strong themes which occur throughout the collection, and we don't think any reader would have expected otherwise, are the varied ways in which language is used to characterise, or pin down 'care', caring and various delineations of these terms. In our opening to this book, we discussed our belief that using care as a lens through which to re-evaluate and reframe the ways that we work for and with children and young people might offer positive opportunities to steer away from existing and potentially troublesome conjecture and readily accepted ideas about young people.

The authors of this collection have gone far in pushing us to think harder, showing just how challenging it is to position ourselves as researchers, policy makers, institutions, service providers or practitioners or in relation to these ideas. Hughes (Chapter 2), for example, adeptly argues that an understanding of 'care' is often assumed, however it remains an ambiguous term which remains elusive to define. Others show (implicitly or explicitly) the relational and reciprocal nature of care and the complexities associated with its constantly changing nature (Hayes, Chapter 11). Perhaps then, as all of the types of individuals above (researchers, policy-makers, etc.) who seek to care for, with or about children and young people, our focus should be less on identifying a categorical definition of care, and more on reflecting on how we position ourselves and our settings or provisions in relation to ensuring that young people can be and feel cared for, and experience the emotional aspects of care through caring about (see our introductory discussion) people, places, issues and outcomes.

What do we mean when we talk about the practice of care? We don't mean that we should only consider the formally recognised requirements and expectations; the everyday, the mundane, are just as powerful. A person in a position of care toward a child may meet all of the standards and targets set for them at a legislative and professional level, but may still not been deemed a caring individual, by peers and by children and young people themselves. We need to reconsider how we frame care for these people, how we can offer support and evidence to show the impact that the everyday, seemingly mundane experiences of caring can have. Further, as a society/organisation/practitioner/individual we need to consider how we can make space for those people who work with children and young people directly to care and, crucially, to show that they care as they carry out their work.

Care and relationships

Who are we talking about when we talk about practitioners and service providers who care? This reflection is written for those who work with children and young people through a variety of contexts, professions and agendas, seeking to enhance the support and care that they offer to and for young people. This collection has offered us a myriad of positive examples of the ways that practitioners can and do care about and for children and young people. These accounts have crossed professional (e.g. education to social care), geographical (we have contributions from the UK, New Zealand [Wood], British Columbia [de Leeuw], Australia [Robinson et al.] and Zambia [Day]) and generational boundaries, extending our understandings of the reciprocal nature of care and resituating children and young people as givers and receivers of the same.

The ever changing, relational aspects of care are a challenging terrain to navigate. None more so than for practitioners working with children and young people. A number of the chapters in this collection signal the ways in which services and professionals can and do adapt to the changing care needs of children and families (this point is made explicitly by Pratchett and Rees in Chapter 3). Day (Chapter 9) argues that children and young people are part of a continuum of care which reflects this diversity, signalling how drawing a distinction between caregiver or care-receiver within this framework is unhelpful (see also Evans' work in Chapter 10). For practitioners and services then, thinking of care in this way highlights a need for flexibility in adjusting input as necessary. Further, Day's chapter illustrates the challenges – in every context – in ensuring that young people who care for others are supported and cared for themselves. This is echoed in Wilkinson's contribution (Chapter 12) on the caring practices of young people who care for intoxicated friends.

Sometimes, caring means simply *doing nothing*. We are reminded of this by Pairman and Dalli (Chapter 8) who show that sometimes children and young people need a timeout (and not in the disciplinary sense) in order to be able to control their levels of social interaction. Hayes (Chapter 11) offers us one way in which we can approach this challenge, by reminding us of the importance of self-awareness in care/caring. Her work illustrates the importance of *showing* that we care rather than going through the motions in providing care. Only when we are self-aware of our comfort zones and our own expectations can we hope to gain a greater understanding of how and why we care, and indeed how this is reflected in the actions of others.

We (care stakeholders) also need to consider how we can show that we care about those who are in caring roles with children and young people. We cannot expect the (emotional) reserves of those who care for and about children and young people to provide for their caring ad infinitum; we need to ensure that they are cared for too. The contributions in this collection have shown the broad and challenging issues that practitioners face on a daily basis (for example caring about and for young people who engage in risky behaviours including self-harm and supporting those who are no longer able to live with their families [and indeed some who may no longer be in contact with them]). These challenges are not without reward, however in order to ensure that they have the capacity to care (about young people, position, profession, organisation etc.) they must be adequately supported.

Care, practitioners, service providers and place

Just as those practitioners who care for children may not, in themselves, be considered 'caring', so too can spaces of 'care' limit caring practices, posing yet another challenge for the practitioners who work within them. Disney (Chapter 6) reminds us of this in his discussion of orphanages which are simultaneously designed to care for children and to segregate them from the outside world. In Chapter 7, Schliehe and Crowley's discussion on young people in secure care has sensitively signalled the challenges which arise for practitioners where a constant balancing act is in play. In these spaces practitioners work and care within formal care spaces which are designed to keep risk at a minimum, whilst also offering a caring environment for the young people that they care for. In the words of the authors these spaces (and the practitioners within them) employ 'control in order to care'. The chapter illustrates how caring through a framework of control is viewed by some young people as preferable to a contrasting lack of stability outside these spaces, to such a degree that they 'engineer' to return. This poses a stark reminder of the importance of recognising the ongoing-ness of children and young people's lives (Horton and Kraftl 2006). If as a society/service/organisation/individual we really care *about* children and young people, we need to work towards provision of not only caring for now, but also caring for the *later* (days, weeks, months, years) *that becomes now* for the children and young people concerned.

Spaces/places and importantly the presence (or not) of practitioners in those spaces can afford (or not) children and young people a vast array of opportunities to participate in or instigate caring acts, and to care about the things and people around them. Wood and Taylor (Chapter 5) demonstrate these reflections, showing how with targeted input, professionals can spark a sense of social (in) justice in young people, motivating them to participate in – and care about – social/community issues. The latter point is also wonderfully illustrated in Hayes' chapter on young people's interactions with nature and the (almost-unnoticed-but-for-a-researcher) involuntary 'wow' that represented a crucial moment of change in appreciation of surroundings.

Pairman and Dalli's work signifies how younger children can appropriate 'potential space' when it is available, enabling them to take control, co-constructing their own care and regulating their levels of social interaction (affording 'downtime' when it is needed). Environments, practitioners and therefore settings (including policy-makers and commissioners) are crucial to affording these opportunities. Practitioners also have a key role to play in affording these instances; we have seen through the contributions here the positive impact that their input can offer where they are given appropriate spaces and scope.

Evidencing care/caring

The words 'evidence-based' (EB) or evidence-based practice are used to describe a myriad of activities in, for example, health, social care and education. The underlying principles of EB in these disciplines hold the same essence, ensuring that where decisions are made and actions are taken, they are based on reliable evidence. As a starting point we want to reflect here on Sackett et al's (1996, p.71) assertion that "Evidence-based medicine is the conscientious explicit and judicious use of current best evidence in making decisions [about the care of individual patients]". Used widely in medical and social care teachings on the importance of evidence, this statement, whilst arising from a medical context, is useful for practitioners of a range of specialisms. Practitioners should seek to identify the current best evidence on which to base their practice. Note the terminology here: current best evidence.

The authors here have presented us evidence drawn from a range of methods, from audits of service-level data to participatory work with young people including storytelling and self-directed photography. Their insightful contributions have explored both macro-scale trends and the intricacies of everyday experiences and emotions arising from the worlds of young people, and those who care for, with and about them. In essence this shows us that evidence for practice is not only that which is undertaken as part of research; evidence in the form of practitioner experience and gut instincts also have a vital role to play in ensuring the place of 'care' in the lives of young people. Perhaps we too readily prioritise evidence for policy development, service planning and delivery in the form of research, audit or evaluation (of existing provision). Readily available evidence through which we can redefine or reframe young people and their lifeworlds is also present in our everyday experiences as is evidenced time and time again throughout this volume. Evidence comes in a range of shapes and sizes; we just need to be ready to take a step back and look for it.

Authors in the previous chapters (and elsewhere) have argued the need for us to find more compelling ways to ensure that where evidence is gathered in the form of consultation, participatory practice, research, audit or evaluation it has the means to have (applied) impact. The term impact will conjure different meanings to practitioners in diverse specialisms, settings and roles. The challenge is set then, to ensure that as stakeholders to care/caring, we reflect on what this means for us, the young people that we care with, for or about, our organisations and for policy or commissioning. We can strengthen our potential for impact by defining clear pathways for dissemination, working together with other care stakeholders to identify common values and priorities and by offering clarity when we identify where the collection/collation/dissemination of evidence will be of use (see the influential works of Kirby et al 2003, Shier 2001, Sinclair 2004 for example). As one example from this collection (for the sake of brevity), Wells' work (Chapter 14) provides us with an excellent account of research conducted based on strong partnership working between charitable organisations, social care and academia, and producing a thought-provoking and sensitive interpretation of the accounts offered by children who are privately fostered across international borders.

A series of questions ... and a challenge

The reflections above lead us to a series of questions that may form the basis of useful reflections for practitioners working with children and young people, and attempting to work in more caring ways with children and young people. These are intended as a starting point for discussions through which we/you/others can spark reflections and action for a reconsideration of care [read caring with, caring for, caring about] children and young people:

1. How is 'care' defined by:
   a. and within the values of your setting
   b. your colleagues
   c. the children and young people that you work with
2. How is care played out/given/received in your particular setting/context?
3. Who holds what role in the everyday caring situations that you experience, when 'care' is offered or received, or when those you work/play/encounter care about or care for others?
4. What affordances do the environments that the children/young people that you practice with, or commission, offer their users opportunities to care with, for and about the contexts around them?

Lastly, we want to leave this reflection with a challenge made to you, the reader. The contributors within this book, and indeed those participating in the conference session from which this collection was developed, all advocate the importance of working with children and young people, developing opportunities to hear their views. We feel strongly that working with young people to understand how they identify with the term 'care' and their needs in relation to the term is fundamental in offering practitioners, providers, commissioners and other individuals who care about, for and with young people opportunities to get closer to their lifeworlds. Our challenge to you is: identify an opportunity to raise the topic of care with a child or young person, listen and hear what they say, and reflect on their response in relation to your own practice.

References

Horton, J. and Kraftl, P. (2006) Not just growing up, but *going on*: materials, spacings, bodies, situations. *Children's Geographies* 4, pp.259–76.

Kirby, P., Lanyon, C., Cronin, K. and Sinclair, R. (2003) Building a culture of participation: involving children and young people in policy, service planning, delivery and evaluation. London, National Children's Bureau.

Sackett, D., Rosenberg, W., Gray, Haynes, B.M. and Richardson, S. (1996) Evidence based medicine: what it is and what it isn't. *British Medical Journal* 312, pp.71–2.

Shier, H. (2001) Pathways to participation: openings, opportunities and obligations. *Children and Society* 15, pp.107–17.

Sinclair, R. (2004) Participation in practice: making it meaningful, effective and sustainable. *Children and Society* 18, pp.106–18.

17 Reflection 3

Children, young people and 'caringscapes'

Sophie Bowlby

The chapters in this book illustrate clearly the diversity of situations in which care for and by children and young people is a dominant focus. They also discuss the diversity of understandings of 'care' that can be at play in such situations. In this contribution I want first to reflect on the complexity and difficulty of identifying a 'caring practice' by discussing two terms often used in descriptions of 'care' that I think are of particular relevance to children and young people and which were highlighted for me by reading the book. Secondly I want to reflect on how a 'caringscape' perspective might be useful to studies of children, young people and care.

First, then, the terms *control* and *obligation* are often used when describing inadequate care – care which does not stem from a proper emotional concern with and focus on the needs for those to be cared for (Tronto 1993).

Control

Providing care to someone may become an opportunity for physical or psychological control, confining the activities and self-expression of the cared-for person spatially, temporally or psychologically or giving inappropriate care. Such abuse of control is associated with failing to listen to or respect the wishes and opinions of the person cared for and was one element of the campaigns of disability activists to take control over the 'care' or assistance they needed (Oliver 1990, Morris 1993). There is an obvious parallel with the concerns expressed by many authors about the need to listen to the voices of children and young people, to recognise their agency and to provide situations in which children can exercise their creativity and make decisions and choices for themselves. True care for children, from this point of view, requires the minimum of control. Yet it is also argued by many that some degree of 'control' over the activities of children and young people is 'necessary' for 'their own good' or for the good of the group as a whole – for example, to 'make them' go to bed early enough to get adequate sleep; to provide 'clear boundaries' for 'acceptable' behaviour; to 'teach them' how to 'eat healthy'. In such cases it seems clear to those exerting control that they know what is 'for the good' of the children and young people they are caring for. These forms of control are seen by their proponents as examples of caring behaviours that adults 'owe' to the children and young people in their care. Fierce debates over such issues and the appropriate balance between necessary control of children and young people's behaviours and enabling their agency can be found in both popular and academic media over many years.

Several chapters in the book explore the complexities of combining care with control, for example, the chapters by Schliehe and Crowley and Disney and the difficulties of defining good care (the chapters by de Leeuw, Hughes and Hayes). They illustrate that there is no general answer to the question of what control is needed for the proper care of children and young people. The answers given in practice depend on the particularities of the situation. These particularities include assessments of the mental and physical maturity of the child or young person (Valentine 2003, 2007) which vary by class and culture. Thus, ideas about the necessarily changing and embodied, but socially shaped, nature of children's mental and physical capacities, and views of what 'personhood' entails and when it comes to exist for a child, are culturally highly variable – especially between the minority (or WEIRD, i.e. 'Western, educated, industrialised, rich and democratic') world and the majority world – as are ideas about the type of 'caring/control' that parents and others should engage in, and what types of independence should be encouraged (Valentine 2007, Otto and Keller 2014). Awareness of the cultural specificity of our understandings of childhood and young people and of care itself is of vital importance to all studies of care and, in particular, to study of care in the lives of children and young people (Raghuram 2016). This will not be news to the authors or most of the readers of this book. Nevertheless, I want to signal the importance of recognising cultural variability and the significance of children's changing, and socially shaped, embodied capacities for destabilising taken-for-granted ideas of children and young people and argue that we should consider their implications for the form and consequences of social policies.

Obligation

In the literature on care it is widely recognised that relationships of care are relational, usually involve reciprocation and that ideas of who should care are socially constructed and vary over time and space. Indeed, much of feminist writing on care and on Ethics of Care has been devoted to challenging the notion that women are 'naturally selected' for caring roles and hence more 'obliged' to care than men. Feminist writing on care has argued for the significance of care to human flourishing and urged that enabling and positively valuing care and caring, financially as well as morally, should be a central organising principle in society (Tronto 1993, 2010, Sevenhuijsen 1998, Skeggs 2014).

It is not only women who can feel obliged to care. Children, friends and family members of either gender may feel obliged to care, as can be seen in the chapters by Pratchett and Rees, Day, Evans, Wells and Wilkinson. Feeling an obligation to care can reflect both the pressures for group survival – as in the inter-generational contract – as well as individuals' sense of group belonging – as in obligations to care for members of the family (however defined) – and individual histories of caring exchanges – as in friendships – and the moral narratives propounded by the state, religious and other groups (Evans 2014, Brannen 2006, Finch and Mason 1991). Thus, the obligations of children and young people to care for others in their family or beyond or for 'the environment', and the obligations of parents, others, or the state to care for children, vary with the social context in which they learn who they are and what socially appropriate behaviour is required of them.

On the one hand it is recognised that feelings of obligation to care are socially desirable and flow from our relations with others in time-space. On the other, there is often an implication that caring that stems from a feeling of obligation may lead to grudging or half-hearted care relationships.

Alternatively, obligations to care may be seen as stunting the lives of both old and young people and limiting opportunity. Obligations created through paying people to care create particular anxieties, especially if that care is to be delivered in the space of the home, which is often presented as a space of love rather than money. We want those delivering care to our children – teachers, nannies, care workers – to demonstrate vocation, love, and empathy and we want our children and young people to care for others, or for nature, out of a sense of genuine concern and love and not simply out of duty. Yet, if caring is to be properly socially valued, provided and supported, we cannot be satisfied by expecting care work to be done for no reward other than personal satisfaction (Skeggs 2014). A positive feminist politics that seeks to re-validate care and caring must find ways to encourage everyone to make an emotional commitment to exchanging care with others and so feel 'obliged' to care. Geographical research on the development of caring relationships in the lives of children and youth can help contribute to this endeavour.

Children's caringscapes

The framework of caringscapes was developed out of an interest in the problems for women of combining paid work with the domestic labour of caring for a spouse, children and/or elderly relatives and in the policies related to them (McKie et al 2002). We wanted to include recognition of the time-space demands of the mobilities and practices of caring with the emotional demands of organising care. We moved on to think of how the intersection of multiple temporalities (e.g. body-time, career-time, employment-time) and spatialities (e.g. places, mobilities, scales, boundaries) might be drawn into a person's caring choices – or lack of choices – over the lifecourse (see Chapter 1). We also suggested that the individual's 'caringscape' had to be seen as interacting with the changing 'carescape' of resources relevant to caring – which might include transport, housing and other local infrastructures not explicitly linked to caregiving – provided through public and private institutions, and regulated by governments (Bowlby et al 2010, Bowlby 2012).

The caringscapes framework provided an adult-centred view of caregiving in which children were seen primarily as recipients of care (see also Robinson et al, Chapter 13) although we did suggest relationships with children could be sources of joy and pleasure as well as of obligation. But children and young people have caringscapes and carescapes also. It is worth asking how placing children or young people as the subjects of a caringscape/carescape study might change the ways in which the time-space relationships of their caring exchanges are viewed.

First, we can consider how some of the issues raised above about control and obligation might be seen from the viewpoint of a child or young person. Clearly, children or young people may experience adult control as a mixture of care and restriction and may themselves seek to exert caring control over others of importance to them. As studies in this book and elsewhere indicate, control of spaces of privacy or of friendship in real and virtual space over a variety of timescales can be important to children's current and future relationships with others and to self-care while enclosure can constrain contact with others or nature (chapters by Pairman and Dalli and Hayes). Obligations to peers loom large in the lives of children and are maintained through networks of contacts at school, home and in virtual space. How these affect their everyday caring practices and ideas of future caring relationships are issues highlighted by a caringscapes perspective.

Secondly, the imaginaries and uses of time-space will alter with the changing, socially shaped, embodied capacities of the child or young person, their experiences of different environments and the people they interact with in these environments. Some of these issues are touched upon in the chapters by Pairman and Dalli and Wood and Taylor. These imaginaries and uses of time-space also will be culturally variable. Caringscapes lay a particular emphasis upon the ways in which memories of past relationships and practices may influence the present and anticipated future as is discussed in children's reactions to parental death in Evans' chapter. Thinking about the caringscapes of children suggests we should explore what kinds of futures children and young people in different socio-spatial contexts and with different embodied social capital (Holt 2008, 2010) imagine and how they enrol past experiences into present and anticipated caring relationships.

Thirdly, we might consider what are the relevant carescapes for young people and children. What are the particular resources of importance to them, how and by whom are they being provided and how and why is that provision changing? The significant resources in the eyes of children and young people may or may not be those that are obviously provided for them. Examining change in resource availability is of particular importance to people in the minority world at a time of continuing austerity and to those in the majority world who already lack many resources for education and health and who may be seeing changes in family and community supports with continuing urbanisation and economic shifts or as a result of conflict. For both, understanding the complex political interactions within and between spatial scales leading to shifts in the availability of resources in particular places remains of great importance.

One aim of the caringscapes/carescapes framework has been to suggest questions that researchers or policy-makers might pose to themselves when assessing particular issues involving care (Bowlby et al 2010). In particular, each of the eight propositions about caringscapes that we suggested (see Chapter 1) were intended to lead to a range of questions that might be asked in a wide variety of situations. I hope that researchers and policy-makers interested in geographies of children and youth will find it worth exploring these questions and using them to cast a critical eye on policy.

References

Bowlby, S.R., McKie, L., Gregory, S. and MacPherson, I. (2010) *Interdependency and care over the lifecourse*, London, Routledge.

Bowlby S. (2012) Recognising the time–space dimensions of care: caringscapes and carescapes. *Environment and Planning A* 44, pp.2101–18

Brannen, J. (2006) Cultures of intergenerational transmission in four-generation families. *Sociological Review* 54, pp.133–54.

Evans, R. (2014) Parental death as a vital conjuncture? Intergenerational care and responsibility following bereavement in Senegal. *Social and Cultural Geography* 15(5), pp. 547–570.

Finch, J. and Mason, J. (1991) Obligations of kinship in contemporary Britain: is there normative agreement? *British Journal of Sociology* 42, pp.345–67.

Holt, L. (2008) Embodied social capital and geographic perspectives: performing the habitus. *Progress in Human Geography* 32, pp. 227–46.

Holt, L. (2010) Young people's embodied social capital and performing disability. *Children's Geographies* 8, pp.25–37.

McKie, L., Gregory, S. and Bowlby, S. (2002) Shadow times: the temporal and spatial frameworks and experiences of caring and working. *Sociology* 36, pp.897–924.

Morris, J. (1993) *Independent lives? Community care and disabled people*. London, Macmillan.

Oliver, M. (1990) *The politics of disablement*. Basingstoke, MacMillan.

Otto, H. and Keller, H. (2014) (eds.) *Different faces of attachment: cultural variations on a universal human need*. Cambridge, Cambridge University Press.

Raghuram, P. (2016) *Locating care ethics beyond the Global North*. Available at: http://ethicsofcare.org/wp-content/uploads/2016/04/Parvati-Raghuram-DISLOCATING-CARE-ETHICS-PLACE-AND-RELATIONALITY-IN-A-GLOBALISING-WORLD.pdf

Sevenhuijsen, S. (1998) *Citizenship and the ethics of care: feminist considerations on justice, morality and politics*. London, Routledge.

Skeggs, B. (2014) Values beyond value? Is anything beyond the logic of capital? *British Journal of Sociology* 65, pp.1–20.

Tronto, J. (1993) *Moral boundaries: a political argument for an ethic of care*. London, Routledge.

Tronto, J. (2010) Creating caring institutions: politics, plurality and purpose. *Ethics and social welfare* 4, pp.158–71.

Valentine, G. (2003) Boundary crossings: transitions from childhood to adulthood. *Children's Geographies* 1, pp.37–52

Valentine, G. (2007) 'My son's a bit dizzy'. 'My wife's a bit soft': gender, children and cultures of parenting. *Gender, Place and Culture* 4, pp.37–2.

Index

For Product Safety Concerns and Information please contact our EU
representative GPSR@taylorandfrancis.com
Taylor & Francis Verlag GmbH, Kaufingerstraße 24, 80331 München, Germany

www.ingramcontent.com/pod-product-compliance
Ingram Content Group UK Ltd.
Pitfield, Milton Keynes, MK11 3LW, UK
UKHW020012200926
475702UK00014B/1849

* 9 7 8 0 3 6 7 2 1 8 8 4 3 *